The
Sec[ret]
Life of
EQUATIONS

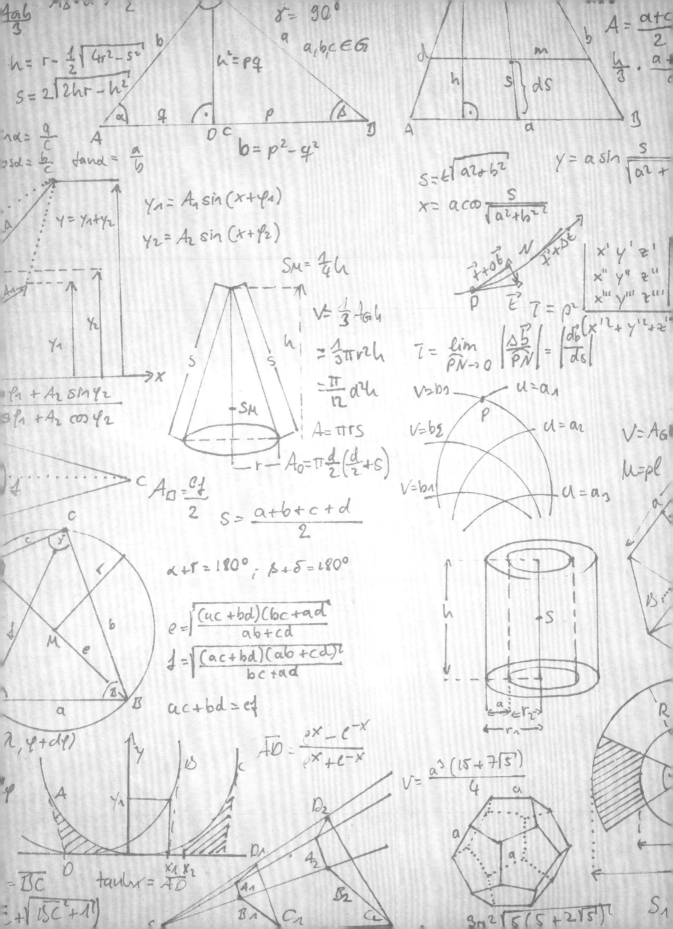

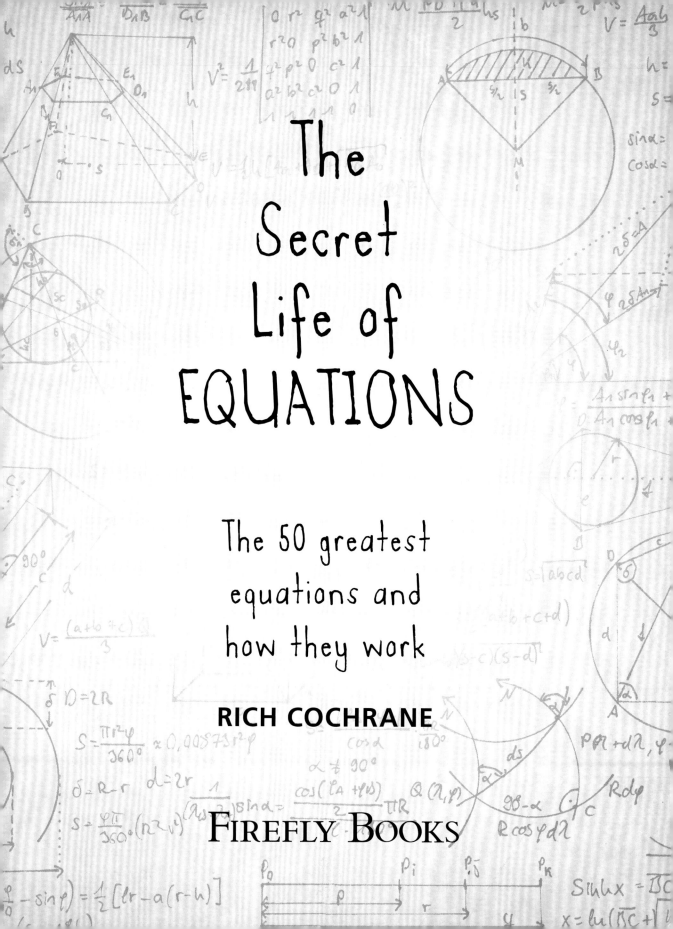

The Secret Life of EQUATIONS

The 50 greatest equations and how they work

RICH COCHRANE

FIREFLY BOOKS

A FIREFLY BOOK

Published by Firefly Books Ltd. 2016

First printing

Publisher Cataloging-in-Publication Data (U.S.)

Names: Cochrane, Rich, author. | Davies, Trevor, editor.
Title: The secret life of equations : the 50 greatest equations and
    how they work / Rich Cochrane, (Editor) Trevor Davies.
    Description: Richmond Hill, Ontario, Canada : Firefly Books, 2016.
    | Includes index. | Summary: "Discover the 50 equations that have
    led to incredible discoveries, ground-breaking technology and
    have shaped our understanding of the world" — Provided by
    publisher.
Identifiers: ISBN 978-1-77085-808-4 (paperback)
Subjects: LCSH: Equations.
Classification: LCC QA211.C634 |DDC 512.94 – dc23

Library and Archives Canada Cataloguing in Publication

Cochrane, Richard (Richard John), author
        The secret life of equations : the 50 greatest equations and
how they work / Rich Cochrane ; Trevor Davies, editor.
Includes index.
ISBN 978-1-77085-808-4 (paperback)
        1. Equations—Popular works. 2. Mathematics—History—
Popular works. I. Title.
QA211.C64 2016   512.9'4   C2016-900567-4

Published in the United States by   Published in Canada by
Firefly Books (U.S.) Inc.   Firefly Books Ltd.
P.O. Box 1338, Ellicott Station   50 Staples Avenue, Unit 1
Buffalo, New York 14205   Richmond Hill, Ontario L4B 0A7

Printed and bound in China

First published by Cassell, a division of Octopus
Publishing Group Ltd
Carmelite House
50 Victoria Embankment
London EC4Y 0DZ
Richard Cochrane asserts the moral right to be
identified as the author of this work.

Contents

Introduction

Some time around 820 CE the Persian mathematician Abu 'Abdallah Muhammad ibn Musa al-Khwarizmi wrote his *Compendious Book on Calculation by Completion and Balancing*. There, he gave us the word "algebra" and gathered together some of its basic principles. Fundamental to algebra is the notion of balance, which the equation has come to embody: if we put an apple on one side and an orange on the other, the scales balance when the two weights are equal. And that's what every equation says: *these two things balance*.

How to use this book

This book could be read from cover to cover, like a novel, but most people don't read math books that way. Math is a vast, interconnected network of ideas that asks to be explored rather than watching it whizz by in a predefined order. For that reason this book contains many cross-references and you may well find that one section makes more sense when you come back to it after reading another, later one. Don't be dismayed by this; it's how most of us feel most of the time when studying math. Even great mathematicians sometimes report feeling lost and confused when learning a new area of the subject. They also describe the joy of finding unexpected connections, some of which can be very profound and beautiful.

When meeting any new mathematical idea, most of us need to start with an intuitive picture of what's going on. This book can't get too technical — every one of its sections has been the subject of whole books, sometimes many advanced and complicated ones. What it can do is make the general ideas plain and indicate ways in which those ideas can talk to each other, sometimes across widely different parts of mathematics, science and everyday life. This inevitably involves some quite drastic simplifications, which I hope beginners will appreciate and experts will forgive. For similar reasons most of the graphs don't have numerical scales and whatnot; I suppose this will enrage math teachers, but the removal of extraneous details keeps our focus on the overall shape of what's going on.

Notation

Still, this is a book about equations. Many popular math books carefully chart their route to avoid too many scary-looking formulas. This one takes the opposite approach. Mathematicians' notation is designed to make life easier, not more difficult. In this respect it's just like other special forms of notation, such as that used by musicians, editors, choreographers, knitters and chess-players. If you can't parse it, it's completely incomprehensible. But if you can, then, like a picture, this notation can do the work of many cumbersome words.

Our way of writing math down isn't always logical. It developed over hundreds of years and it can be quirky, weird or downright silly; like most things it shows traces of the historical process that produced it. Perhaps someone could invent a whole new way of writing equations that would be more coherent, but only a foolhardy reformer would dare to try. So don't worry if sometimes it's clear to you what a symbol represents but not why it looks the way it does. At some point you learned to read the words on this page; that was a much more difficult task, involving a system of notation that's almost completely arbitrary. If you managed that, you can surely do this too.

I've assumed you know about positive and negative whole numbers and what a fraction is, along with the following principles from algebra. Letters (or other symbols) can be used to stand for numbers that are unknown or that can vary. Multiplying these unknown quantities can be

Table of Symbols

Here's a list of the most important symbols that crop up in multiple sections, along with the place where they're first introduced.

$\sqrt{x}$	Square root of x [Pythagoras's Theorem, page 10]	x', x''	First and second derivatives of x with respect to time (alternative notation) [Curvature, page 30]
Σ	Sum [Zeno's Dichotomy, page 18]	log, ln	Logarithms [Logarithms, page 36]
lim	Limit [Zeno's Dichotomy, page 18]	i	The square root of −1 [Euler's Identity, page 40]
∞	Infinity [Zeno's Dichotomy, page 18]	∇^2	Laplacian [The Heat Equation, page 80]
π	Pi [Euler's Identity, page 40]	div, curl	Derivatives of vector fields [Maxwell's Equations, page 92]
sin, cos, tan	Trigonometric functions [Trigonometry, page 14]	∇	Gradient [The Navier-Stokes Equation, page 96]
$\int$	Integral [The Fundamental Theorem of Calculus, page 26]	$\neg$, $\wedge$, $\vee$	Logical not, and, or [De Morgan's Laws, page 126]
$\dfrac{dy}{dx}$	Derivative of y with respect to x (other letters sometimes replace y and x) [The Fundamental Theorem of Calculus, page 26]	$P(x)$	The probability of event x [The Uniform Distribution, page 162]
$\dfrac{d^2y}{dx^2}$	Second derivative of y with respect to x (other letters sometimes replace y and x) [The Fundamental Theorem of Calculus, page 26]	$P(x\|y)$	The conditional probability of x given y [Bayes's Theorem, page 168]

represented by putting the letters next to each other, so that

$$a \times b = a\,b$$

and dividing one number by another is conveniently represented by a fraction:

$$a \div b = \frac{a}{b}$$

Finally, the all-important equals sign says that the total of everything on one side of it is exactly the same as on the other. All other notation will be introduced as we go along.

Each equation works like a little machine with moving parts; our main task is to understand what each part does and how it interacts with all the others. Sometimes that means spending time unpacking or decoding notation. Sometimes it means working through a simple example. Sometimes it means getting to the bottom of

something obscure or, by contrast, catching a hurried glimpse of it as we zoom past.

In fact, in terms of a traditional sequence of math education, this book is incredibly uneven: one minute you're dealing with a bit of high-school algebra, then on the next page you hit something you'd meet only much later in a university degree. I've chosen to ignore that, because mathematical subjects don't come with predefined levels of difficulty. The arithmetic you learned as a child is, it turns out, incredibly deep and mysterious, while many so-called "advanced" topics are actually pretty easy to grasp once you get past the jargon. Go with the flow, understand what you can and look more deeply into the parts that grab your interest. There isn't a wrong way to do this.

Introduction

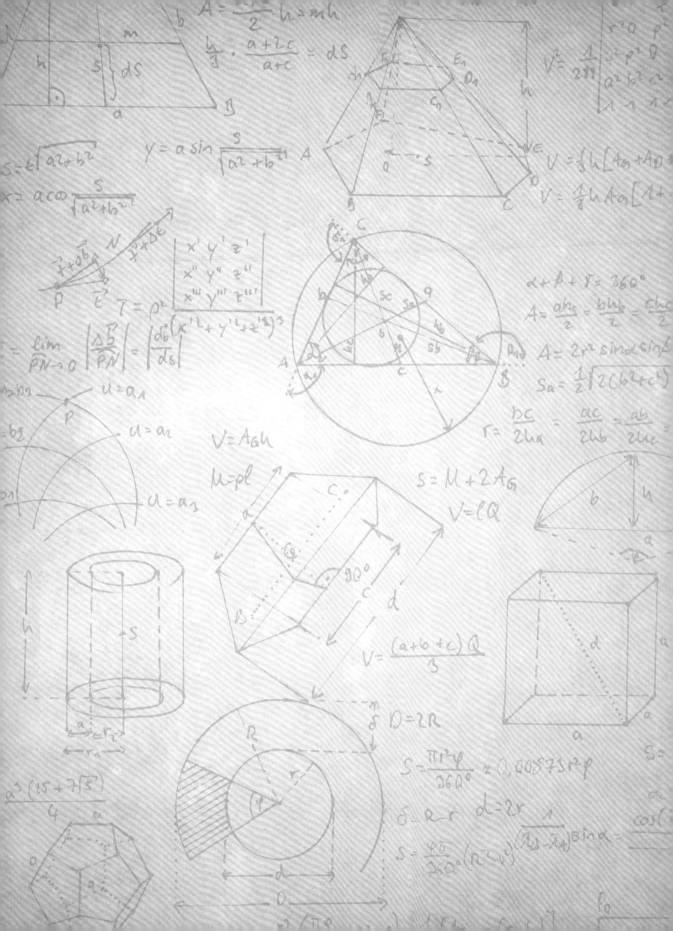

The Shape of Space

GEOMETRY AND NUMBER

Pythagoras's Theorem

The sides of a triangle tell us something basic about how space works.

The long side

The other sides

$$A^2 = B^2 + C^2$$

What's It About?

Take any three sticks of any lengths. Call the lengths *A*, *B* and *C*, and suppose *A* is the longest one (or joint-longest one, if necessary). You'll find you can make the sticks into a triangle as long as *A* is less than *B* + *C*. If you want to make a triangle with a right angle — a 90° corner, like the kind on a square or rectangle — you have to have a very special set of sticks. In fact, if you have *any* two *B* and *C* sticks already fixed in a right angle (making an L-shape), Pythagoras's Theorem tells you how long stick *A* must be to complete the triangle.

At first this might seem less than impressive. First, it only works for a triangle with a right angle in it, which seems a bit of a limitation. Second, when *was* the last time you had to work out the lengths of the sides of a triangle anyway? Well, it turns out that triangles are fantastically important. In a sense a triangle is the simplest two-dimensional shape, you can make, so problems involving other 2D shapes can often be turned into problems about triangles. Many 3D problems can, too. What's more, right-angled triangles have a rather special place among all their three-sided siblings (see Trigonometry, page 14).

The kind of triangle you get depends on the lengths of its three sides. Some combinations of lengths can't be made into a triangle at all.

Why Does It Matter?

Pythagoras's Theorem is one of the few equations in this book that you might yourself use when, for example, doing a bit of handy work around the house. Still, that doesn't really explain why it's such an important equation. What it captures is something very basic about the way we expect distances to work: in particular, how they relate to the way we judge lengths and distances.

Imagine a big field with a solitary wooden post somewhere around the middle of it. Suppose I've hidden some treasure in a secret location in this field and I need to direct you to the right spot to dig it up by passing you a message that's as concise as possible. As long as I know you'll have a compass with you (or you know how to find north by looking at the sky), I can give you the necessary information using just two numbers. I can tell you to stand at the post and go so many meters (or yards) north, and then so many meters (or yards) east.

What if I've hidden the treasure somewhere south-east of the post? No problem — I can give you a negative number for the north distance, and you'll be able to interpret −10 m (−9.15 yd) north as meaning 10 m (9.15 yd) south. In this way I can identify any point in the field, however large it is, with just those two numbers. In fact this is a standard way to find our way around flat, two-dimensional spaces, and it was formalized by the French mathematician René Descartes in the early 1600s. Instead of north and east we often use *x* and *y*, which you might remember from school. Sometimes physicists use i and j to mean more or less the same thing.

It's not even important where the post is; in fact, if the post moves, I can always adjust the numbers I've given you to take that into account. In a sense, then, this allows us to get from any one point (the post) to any other (the treasure). Here's what Pythagoras does for us: in this setup we know the distances north and east, and these form two sides of a right-angled triangle (because east is at right angles to north). So Pythagoras's Theorem tells us what the direct distance is between the post and the treasure. That makes it a fundamental fact about distances in space.

Perhaps you can see how to extend it to three dimensions, too: simply add another number indicating "height above ground" (see illustration, page 12). If the number is negative, it tells you how deep to dig downwards! Pythagoras's Theorem still works in three dimensions, and even in higher

dimensions, too. We call these setups "rectangular coordinate systems," and Pythagoras's Theorem gives us a way to calculate lengths and distances. This is some of the most basic information we need in math, physics and engineering where these systems — and this equation — are used every day.

In More Detail

We don't know much about Pythagoras; he lived in the Greek world in the fifth century BC and became the leader of a religious cult whose beliefs were steeped in numerology. Many weird stories have been told about his life and teachings, but if he himself ever wrote any of it down, none of it survived. The fact known as Pythagoras's Theorem probably wasn't discovered or proved by him alone, but it certainly seems to have been in circulation among his followers. In the book *The Ascent of Man*, the 20th-century mathematician and author Jacob Bronowski called it "the most important single theorem in all of mathematics"; that might be pushing it a bit, but it's surely one of the ancient mathematicians' great achievements.

The first thing to notice is that on paper this looks like an equation about areas rather than lengths. After all, if *A* is a length, say 10 cm (or 10 in or 10 anything else), then A^2 is the area of a 10 cm × 10 cm square, that is, 100 cm^2 (one

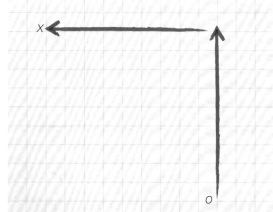

To find the point *X* from the origin *O*, we walk a certain distance up and then to the right. Here, that distance to the right is negative!

hundred square centimeters). That, in fact, is the ancient view of it, summed up in a slogan that schoolchildren down the centuries were made to recite: "The square on the longest side is equal to the sum of the squares on the other two sides." This, though, hardly makes it clear why anyone should care about it, since we very rarely come across three squares arranged so neatly in real life.

The power of the theorem comes from our ability to take square roots. The square root of a number is just the number that, when you multiply it by itself, takes you back to where you started. So the square root of 9 is 3, because 3 × 3 = 9. In other words, if you want to lay out a square room whose area is 9 m^2 you should make each side of the room 3 m long. In modern notation we write

$$\sqrt{9} = 3$$

with that odd tick symbol meaning "square root."

We're now ready to use Pythagoras's Theorem to find the length of stick we need to finish off a triangle or, more excitingly, how far the treasure is from the post. For example, suppose stick B is 3 cm long and stick C is 4 cm; they're already fixed in an L-shape; we want to find the length of stick A to complete the triangle:

$$\begin{aligned} A^2 &= B^2 + C^2 \\ &= 3^2 + 4^2 \\ &= 9 + 16 \\ &= 25 \ \mathbf{cm}^2 \end{aligned}$$

So we know A^2, but we want to find A; that is, we know the area of the square and want the lengths of its sides, which is exactly what the square root gives us:

$$\begin{aligned} A &= \sqrt{25} \\ &= 5 \ \mathbf{cm} \end{aligned}$$

As in the example about areas given above, this sum works equally well with sides of 3 in, 4 in and 5 in, or of any other unit. I didn't pick the numbers 3, 4 and 5 by accident: when A, B and C in Pythagoras's Theorem are all nice whole numbers they're called a "Pythagorean Triple." These aren't

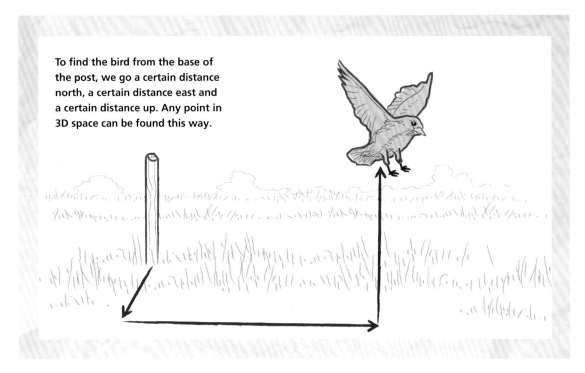

To find the bird from the base of the post, we go a certain distance north, a certain distance east and a certain distance up. Any point in 3D space can be found this way.

The Shape of Space: Geometry and Number

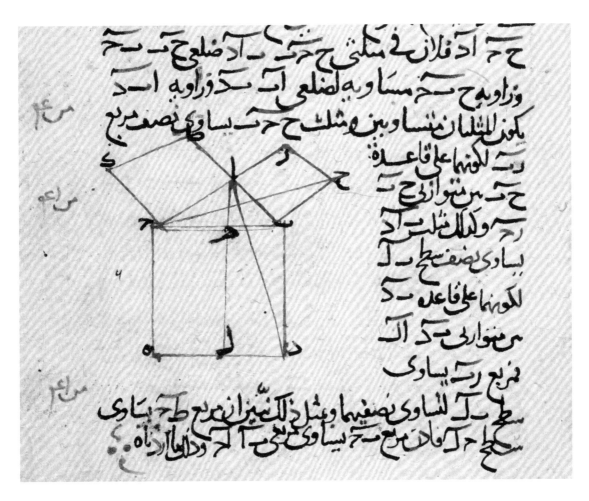

so easy to come up with by trial and error, but the ancient Greek geometer Euclid figured out a clever way to find them. Pick any two different whole numbers — call them p and q — and suppose p is the bigger one. Then make

$$A = p^2 + q^2$$
$$B = 2pq$$
$$C = p^2 - q^2$$

The renowned Persian scholar Nasir al-Din al-Tusi published his version of Euclid's Proof of the Pythagorean theorem in Arabic in 1258.

and you have a Pythagorean Triple. If you know a little bit of algebra, try proving for yourself that this works: $B^2 + C^2$ will always equal A^2 if Euclid's recipe is followed.

Pythagoras's Theorem looks like a relationship between the areas of three squares, but actually it tells us how to work out distances between points in space.

Trigonometry

Circles make the world go round; triangles give us a handle on them.

The angle

$$\sin (a) = \frac{O}{H}$$

The opposite side

The long side

$$\cos (a) = \frac{A}{H}$$

The adjacent side

$$\tan (a) = \frac{O}{A}$$

What's It About?

The word "trigonometry" means something like "the art of measuring triangles." Triangles are some of the most basic shapes in geometry — they come up everywhere in areas such as surveying, building and astronomy, so it's not surprising that this is a very old art indeed. In fact, in some ways trigonometry is older than anything we would recognize as geometry, or even mathematics: we can find its beginnings in practical techniques used in ancient Egypt and Babylon a good 4,000 years ago.

It turns out that trigonometry has intimate connections with circles, even though circles don't look much like triangles. This, too, was known intuitively from very early on: a point moving in a circle can be described by the trigonometric functions, and they appear in many mathematical models that involve circular or smooth back-and-forth motions. As a result, they pop up in several equations in this book.

In More Detail

When it comes to measuring triangles, two things spring to mind: the lengths of the three sides and the sizes of the three angles. These are obviously connected: to see this, take any three sticks and you'll find there's only one triangle you can make with them, so the lengths seem to determine the angles in advance.

This relationship is evidently more about proportions than about actual lengths, though, since two triangles can have the same angles but different-length sides. In other words, they are the same shape but different sizes — the jargon from geometry class is that they're "similar triangles." So it is the ratios of the lengths of the sides that determine the angles in the triangle, not the actual lengths themselves.

Around 600 CE, Indian scholars created the main trigonometric ratios as we know them today,

The Shape of Space: Geometry and Number

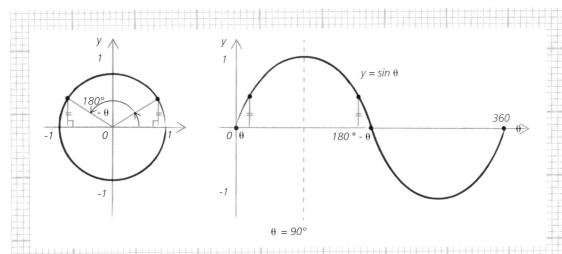

The sine function (on the right) is the height of the green triangle (on the left) as its longest side sweeps around the red circle.

though they went under different names: we call them sine ("sin"), cosine ("cos") and tangent ("tan"). There have been quite a few others, some of which are still in regular use because they make certain formulas or operations more convenient, but these are the best-known ones. For a long time they were laboriously calculated by measuring different triangles. But why would anyone feel the need to do this? The answer, at the time, was simple: trigonometry helps us deal with common, real-life problems that are tough to solve without it.

Suppose you want to measure the height of a tall tree that's too difficult to climb. If you lie down on the ground you can measure the angle you have to look up at to see the top of the tree. This can be done quite accurately with some simple equipment. You can also easily measure the distance along the ground from where you're looking to the bottom of the tree. From this information, trigonometry enables us to find the height of the tree.

We know an angle, x, and the length of side adjacent to it, A; we'd like to find the length of the opposite side, O. The formulas tell us that

$$\tan(x) = \frac{O}{A}$$

Suppose we measure the angle to be 40°. We look up tan(40) in a table — or use a calculator — and find the value is about 0.839. Suppose we also measured the distance A to be 10 m (or, again, 10 yards or 10 of any other unit). Then we have

$$0.839 = \frac{O}{10}$$

which means that O, the height of the tree, must be 8.39 m. As you can imagine, this was a very useful technique for ancient surveyors and builders to know about, and their successors still use it today.

Circles, angles and distances are basic building blocks of some of the math we use most often — trigonometry brings them together in a strange kind of harmony.

Trigonometry

Conic Sections

The circle, ellipse, parabola and hyperbola are found everywhere in nature and have a simple geometric description.

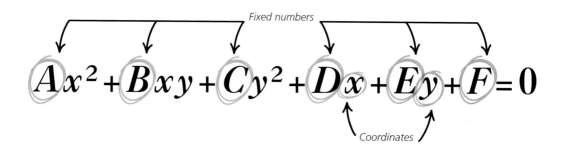

Fixed numbers

$$A x^2 + B x y + C y^2 + D x + E y + F = 0$$

Coordinates

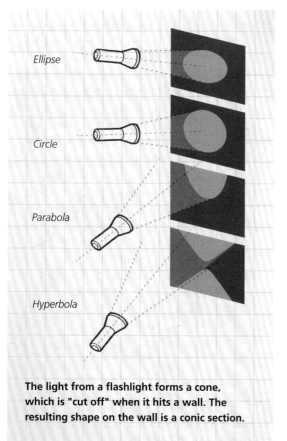

Ellipse

Circle

Parabola

Hyperbola

The light from a flashlight forms a cone, which is "cut off" when it hits a wall. The resulting shape on the wall is a conic section.

What's It About?

Shine a flashlight directly at a wall: it should make a circle of light. Now slowly tilt the flashlight upwards a little bit and watch the circle stretch out into a new shape. If you keep going, at some point the shape will suddenly seem to open out indefinitely as it goes upwards. You can even carry on, and for a while at least you'll see the shape moving upwards and changing shape more subtly. Those shapes, though they look quite different, are all "conic sections." Each one is literally a cross-section of the cone of light coming from the end of the flashlight.

As well as cropping up in many natural settings that seem to have nothing to do with walls and flashlights, conic sections share a surprising geometric unity. This comes from the fact that your flashlight is really producing a constant, three-dimensional cone of light, as you can see if the room is smoky or very dusty. The two-dimensional shape you see changes simply because of the changing angle at which the wall "chops off" the cone.

In More Detail

As you tilt the flashlight the shapes you see are, in sequence, a circle, a series of ellipses, a parabola and then a series of hyperbolas (see illustration).

The Shape of Space: Geometry and Number

Like many power stations, this one in Didcot, England, uses cooling towers whose curved outline is a parabola.

Jets of water often form parabolas, like these at the University of Adelaide, Australia.

These are some of the most important curves in all of mathematics. When you throw a ball, its path is a parabola (see Newton's Second Law, page 56); the same shape is used to make mirrors, microphones and other objects that use reflection to concentrate a signal onto a point, and Archimedes is even said to have used parabolic mirrors to set fire to ships during the Siege of Syracuse in the third century BCE. The planets move in ellipses around the sun (see Kepler's First Law, page 52) and the ellipse has its own reflective properties, exploited in the "whispering galleries" of St. Paul's Cathedral in London and in the treatment of gallstones by sound waves. Hyperbolas can be found in soap films and electrical fields and are frequently used in architecture and design. The image of the flashlight's beam on the wall changes more subtly from parabolic to hyperbolic when the torch is parallel to the wall — for example, when pointing directly upwards — so lamps close to walls usually create hyperbolic shapes.

To draw a curve using the equation given above, first fix values for *A*, *B*, *C*, *D*, *E* and *F*. The other letters (*x* and *y*) define points in a two-dimensional space, so that every point gives a unique pair of values for *x* and *y* (see Pythagoras's Theorem, page 10). Now we try each point to see if the equation is true there — if so, it belongs to the curve, and if not, it doesn't. Most of the points we try won't work; when we calculate everything on the left-hand side of the equals sign, we'll get something other than zero, so that point isn't on the curve. We only choose the points where we get zero, and perhaps we imagine marking them with a dot. What we'll find is that our dots always join up to make one of the shapes the flashlight made on the wall: a circle, ellipse, parabola or hyperbola, depending on the values we choose for our fixed numbers. Actually there are two other possibilities, for if we choose the numbers really carefully we can either get two straight lines that cross each other or just one single point.

The curves known as conic sections fascinated the ancient Greeks and have found a surprising range of applications in the modern world, from lens making to architecture.

Zeno's Dichotomy

A "proof" that motion is impossible comes close to inventing calculus two millennia early.

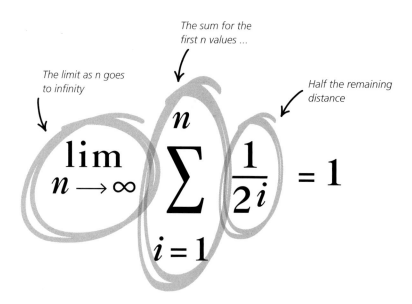

The sum for the first n values …

The limit as n goes to infinity

Half the remaining distance

$$\lim_{n \to \infty} \sum_{i=1}^{n} \frac{1}{2^i} = 1$$

What's It About?

Suppose, says the ancient philosopher Zeno of Elea, that you're in the middle of a room and want to get out. The door is open and there's nothing blocking your path. Go ahead and walk to the door — except there's a tiny problem. To get there you must first walk halfway to the door. Then, you must walk halfway from where you are to the door. You still won't have got there, so you must repeat this again and again … how many times? Zeno thinks the answer is an *infinite* number of times. After all, with each motion you get closer to the door but the next step only covers half of the remaining distance, so you never quite close the gap. Well, he concludes, nobody can do an infinite number of things in a finite amount of time, so getting out of the room is impossible!

Zeno's argument isn't quite as silly as it sounds — as far as we can tell, it was one of a set of four arguments that work together to criticize some specific ancient ideas about space, time and motion — but for us the interest is more mathematical than philosophical. What Zeno has noticed is that a given distance seems to be equal to the sum of all those halves: we halve it, halve it again, halve it again and so on. In modern language he's discovered the idea of a limit, which in the 18th century became a basic tool in math and physics.

Why Does It Matter?

Infinities bother people, and not just in philosophy seminars. The idea that you can add up an infinite number of things and get something perfectly ordinary and finite seems dodgy from the outset; after all, nobody actually could add them all up, since they'd never finish the process. In response to problems like this Aristotle made an important distinction between actual infinities and merely potential ones, which can go on as long as you like without any definite end-point. The easiest

The Shape of Space: Geometry and Number

To get to the door, Betty must first get halfway across the room and then cover half the remaining distance, then half the new remaining distance — a process that seems to go on forever. Can she ever escape?

example is counting: I can count as high as I want, and there's obviously no biggest number (we can always add 1 to get a bigger one!), so counting is a potential infinity, but I can't actually *count to infinity*. What's happening in Zeno's argument looks a lot like "counting to infinity," and that seems decidedly suspicious.

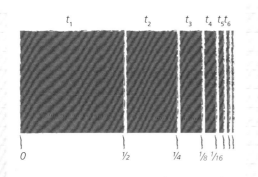

This diagram shows Betty's progress towards the door: each time interval she crosses half the distance remaining. The total distance is finite, but the number of steps is infinite.

This became a major problem when physics started using new methods known as "calculus" at the end of the 17th century. These methods were very useful but they seemed to rely on infinitesimal (infinitely small) distances in ways their inventors couldn't really justify. What if the new physics of Newton and his followers was really based on an absurdity, like Zeno's paradoxical argument? It was very worrying. So in between using calculus to solve mathematical problems and devising successful new physical theories, many people started asking how to make sense of those lurking infinities. The result was the idea of a limit. Though they look fancy, today limits are used widely in math and in most of the applied disciplines that use it.

In More Detail

To unpack what's going on in Zeno's Dichotomy we'll have to understand two standard bits of notation — which, incidentally, Zeno himself wouldn't have understood either. This is worth doing as we'll see them coming up in quite a few other equations in this book, and they're not as scary as they first appear. The first is the big zigzag-like symbol, Σ, which is actually a capital letter sigma from the Greek alphabet. The second is the "lim" itself.

Sigma is the Greek alphabet's equivalent to the letter S, which in this case stands for "sum." Although we can use the word "sum" for any sort of calculation, in this context it specifically means "adding up": whatever comes after the big "sum" sign is going to be added up. But how, exactly?

At the bottom of the sigma is a little equation, "$i = 1$," and at the top is a single letter "n," and these are clues to how to use it. Imagine the sigma is a building with n storeys. We go in at the ground floor ($i = 1$) and start hiking up stairs. Each time we reach a new landing we add 1 to i and then find the value of the thing after the sigma sign. We make a note of the result and then carry on. When we reach the top ($i = n$) we have a list of numbers that we add up (sum) to get the final result.

Here's an example, with 10 storeys to climb:

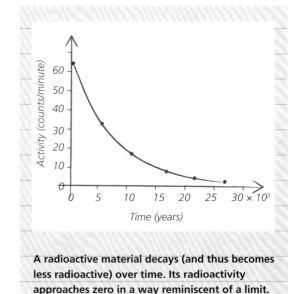

A radioactive material decays (and thus becomes less radioactive) over time. Its radioactivity approaches zero in a way reminiscent of a limit.

$$\sum_{i=1}^{10} \frac{1}{i} = \frac{1}{1} + \frac{1}{2} + \frac{1}{3} + \frac{1}{4} + \frac{1}{5} + \frac{1}{6} + \frac{1}{7} + \frac{1}{8} + \frac{1}{9} + \frac{1}{10}$$

On the left we have a 10-storey sigma, starting with $i = 1$. At each landing we calculate $\frac{1}{i}$, since that's the thing after the sigma sign, and move on. At the end we add them all together to get the answer. You can do this on your calculator if you like, although the result isn't very pretty.

Here's a very similar calculation, which shows how close to the door we are after taking 10 steps in Zeno's Dichotomy:

$$\sum_{i=1}^{10} \frac{1}{2^i} = \frac{1}{2} + \frac{1}{4} + \frac{1}{8} + \frac{1}{16} + \frac{1}{32} + \frac{1}{64} + \frac{1}{128} + \frac{1}{256} + \frac{1}{1,024} + \frac{1}{2,048} = \frac{2,047}{2,048}$$

Take a moment to notice that the sum really does make sense: first we step halfway towards the door, then a quarter of the total distance (half of what's left), then an eighth of the total distance

(half of what's left, again) and so on. Adding these all up gives us a number that tells us we're very close to the door, but not quite there yet.

Nothing forced us to take exactly 10 steps in this example. We can make our notation a bit more general by replacing 10 with n, so you can plug in any number of steps you like and see how close you get to the door:

$$\sum_{i=1}^{n} \frac{1}{2^i}$$

Well, hang on, you might already be saying: Zeno doesn't set us a limit to how many steps we can take to get to the door. He says *no matter how many we take*, though we'll get closer and closer, we'll never quite make it there. In modern terms, we allow n to grow bigger and bigger, without imposing any limit, and we see what happens. This is where the "lim" part comes in.

Let's look at a slightly simpler example:

$$\lim_{n \to \infty} \frac{1}{n}$$

The Shape of Space: Geometry and Number

As *n* gets bigger and bigger, $1/n$ gets smaller and smaller. In fact, it gets awfully close to 0 when *n* is very big. What's more, if you give me any "margin of error," however small, I can find a value of *n* so that $1/n$ is closer to 0 than your margin of error is and, from that point onwards as *n* increases $1/n$ always stays within that margin. In English we say "the limit of $1/n$ as *n* goes to infinity is 0." We don't mean that *n* ever becomes infinity, just that it's allowed to grow larger and larger. That, in a nutshell, is what a limit is.

Let's now look at the claim made by our equation:

$$\lim_{n \to \infty} \sum_{i=1}^{n} \frac{1}{2^i} = 1$$

In English: "the limit, as *n* goes off towards infinity, of the sum of ½ for every *i* from 1 to *n*, is equal to 1." Quite a mouthful, admittedly: but it exactly expresses the geometric intuition that as you take each step covering half the remaining distance you get closer to the door (that is, to have covered a total distance of 1 unit) and that you can get as close as you like to the door if you're allowed to take a lot of steps (though you can never actually get to it).

In fact this sophisticated, 18th-century idea was already almost-formed when Archimedes used his "method of exhaustion" to find the circumference of a circle. He noticed that if you fit regular polygons into a circle, letting the number of sides increase without bounds, they get closer and closer to being circles. In modern terms, Archimedes realized that "the limit of the circumference of a regular polygon, as the number of sides increases, is the circumference of a circle," which gave him an approximate value of pi (see Euler's Identity, page 40).

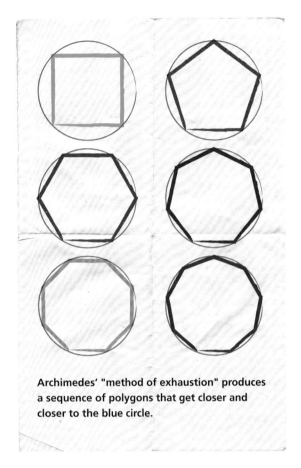

Archimedes' "method of exhaustion" produces a sequence of polygons that get closer and closer to the blue circle.

Approaching a limit by an infinite number of smaller and smaller steps sounds like philosophical wordplay but it lies at the heart of calculus, one of the most useful of all mathematical inventions.

Zeno's Dichotomy

Fibonacci Numbers

What number links pentagons, ancient mysticism and rabbit-breeding?

The next Fibonacci number

The one before it

The one before that

$$F_n = F_{n-1} + F_{n-2}$$

What's It About?

In 1202, Leonardo of Pisa, known as Fibonacci, published and solved the following problem: imagine you're a farmer breeding a special kind of rabbit that reaches sexual maturity at one month old and has a very long lifespan. Each mature female can produce one male and one female each month. You take a new-born male and female and put them in a big field with plenty of food and no predators. Now suppose you let nature take its course and return after a certain number of months, n. How many mating pairs of rabbits will you have? The answer is F_n, the nth Fibonacci number, and our equation tells us how to calculate it.

The problem may seem a bit trivialized: after all, it does not describe a very realistic situation. Yet the Fibonacci numbers are an extraordinary discovery. They have a close relationship with an ancient number known as the Golden Ratio, which many have believed to be sacred or mystical and which itself has many surprising connections to other mathematical puzzles. There are a great many alleged sightings of these numbers in nature, especially in biology, where a simple rule like Fibonacci's might explain how organic growth

that produces complicated-looking forms can be encoded by relatively little DNA.

Why Does It Matter?

The truth is, the Fibonacci numbers have mostly fascinated mathematicians, not scientists or technologists. We'll discuss some mathematical reasons to be interested in them shortly. However, in inventing them, Fibonacci gave birth to the really important general idea of a "recurrence relation." This is, crudely put, any sequence of numbers whose next term depends only on one or more of the terms before it, according to a rule that never changes.

The way recurrence relations evolve the next value from the ones that have gone before makes them very useful for describing processes that develop over time. Extremely simple recurrence relations govern the amount of money in your savings account (assuming you deposit the same amount every month), for example, and the amount you owe on your mortgage (see Logarithms, page 36). Economists often use much more complex recurrence relations than these, as do biologists and engineers. What are known as Markov Chains — essentially, recurrence relations

The Shape of Space: Geometry and Number

that only rely on the immediately preceding value, usually including an element of chance — appear in a perplexing array of applications, from the physics of heat diffusion (see The Heat Equation, page 80) to financial forecasting (see Brownian Motion, page 70).

Some recurrence relations continue to entertain the pure mathematicians, too, and the most notorious is the following. The first item in the sequence is any whole number you choose. The rule is: if the last number was even, halve it; otherwise treble it and add 1. So if we begin with 7 the sequence starts off like this:

7, 22, 11, 34, 17, 52, 26, 13, 40, 20, 10, 5, 16, 8, 4, 2, 1, 4, 2, 1, 4, 2, 1, 4, 2,...

Notice that, after a bit of jumping around, this sequence reaches the number 1, and after that it settles down into a simple cycle of three numbers. Will this always happen, no matter which number you start with? That is, do these sequences always, eventually, hit 1? The so-called Collatz Conjecture says they do; though easy to

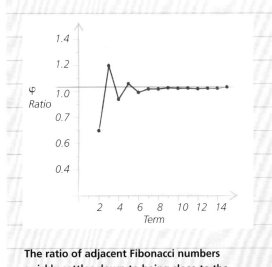

The ratio of adjacent Fibonacci numbers quickly settles down to being close to the Golden Ratio, approaching it as a limit.

understand, nobody knows whether it's true or not. A solution, if it's found, will almost certainly involve the development of brand new ideas, perhaps with wide applications.

The family tree of Fibonacci's rabbits should follow his recurrence relation — at least in theory. Real life is somewhat more complicated.

Fibonacci Numbers

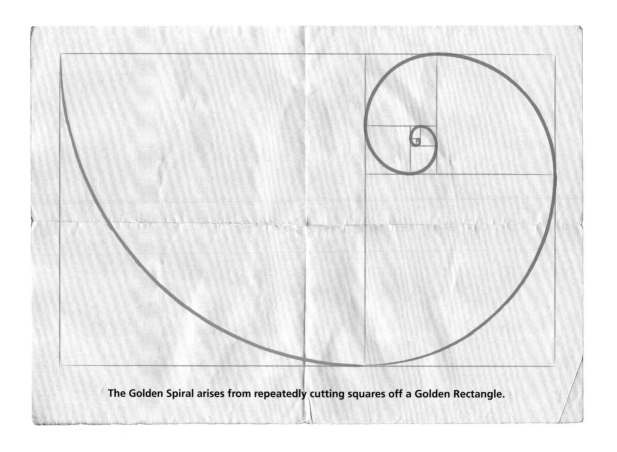

The Golden Spiral arises from repeatedly cutting squares off a Golden Rectangle.

In More Detail

The Fibonacci numbers form a sequence, and if you were to start writing it out F_n is the number you'd find in the nth place on the list. The recipe for making any Fibonacci number is simply to add up the preceding two, but since we need a place to start we set $F_1 = 1$ and $F_2 = 1$. From here we can calculate $F_3 = 1 + 1 = 2$ and we're off. The first few numbers are

$$1, \ 1, \ 2, \ 3, \ 5, \ 8, \ 13, \ 21, \ 34, \ 55,$$
$$89, \ 144, \ 233, \ 377, \ 610, ...$$

and so on forever (or until you get tired of it).

Evidently the numbers in a Fibonacci sequence get bigger and bigger but spotting any other pattern seems a bit tricky. In fact, a lot of patterns are hiding in there.

An important example is what happens to the fraction you get when you divide each Fibonacci number by the previous one:

$$1/1, \ 2/1, \ 3/2, \ 5/3, \ 8/5, \ 13/8,$$
$$21/13, \ 34/21, \ 55/34, \ 89/55, ...$$

Weirdly, if you plug those fractions into a calculator you'll find they get closer and closer together, as if closing in on some specific value. In fact, it can be proved that they approach a limit (see Zeno's Dichotomy, page 18), and that

$$\lim_{n \to \infty} \frac{F_n}{F_{n-1}} = \frac{1 + \sqrt{5}}{2}$$

This limit is usually given the symbol φ (the Greek letter phi). This is the so-called Golden Ratio; it's about (but not exactly) 1.618. This also happens to be just the number you need if you want to

The Shape of Space: Geometry and Number

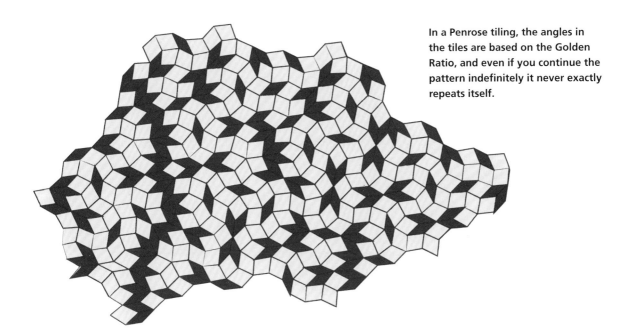

In a Penrose tiling, the angles in the tiles are based on the Golden Ratio, and even if you continue the pattern indefinitely it never exactly repeats itself.

draw a regular pentagon (five-sided shape) using only a ruler and compass; this was an important practical technique for artists and craftsmen and led to the ancient Greeks discovering the ratio, long before Fibonacci and his imaginary rabbits.

The Golden Ratio has an extremely odd history and even today there are people who believe it has something approaching magical powers. They claim it governs many natural phenomena and has been used by architects and artists to create work with intrinsically pleasing proportions. I'm sorry to say that many of these claims turn out to be false. The Golden Ratio does, though, appear in the natural structures known as "quasicrystals," which chemists are still actively researching.

The wilder claims often turn up in relation to the so-called Golden Rectangle. If you cut a square off this type of rectangle, the leftover part has exactly the same proportions as the one you started with. This means you can carry on cutting off smaller and

smaller squares forever (or until you get bored). If you cut up the Golden Rectangle repeatedly, in just the right way, you can draw a quarter-circle in each square and produce a Golden Spiral, which is very pretty. This only works when the rectangle's longest side is φ times its shortest. Let's briefly see why. If it's going to work, the main rectangle will have sides of length, say, 1 and r such that once you've cut a 1×1 square off, the remaining rectangle will have short side equal to $r-1$ and long side equal to 1. So r must satisfy the equation

$$\frac{1}{r} = \frac{r-1}{1}$$

A little rearranging turns this into the equation $r^2 - r - 1 = 0$, and a little high-school algebra yields φ as one of the two possible solutions (the other also works; you just end up with the rectangle flipped over on its side).

Recurrence relations create complex results by repetition, just as many natural processes do, and their long-term behaviors are often full of surprises.

The Fundamental Theorem of Calculus

Calculus is a universal mathematical tool; this equation is what makes it work.

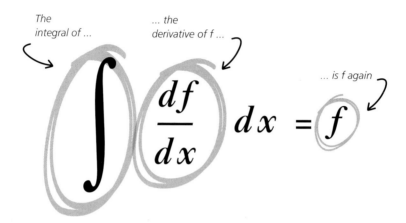

The integral of ...

... the derivative of f ...

... is f again

$$\int \frac{df}{dx}\, dx = f$$

What's It About?

"Calculus" comes from the Latin word for a small stone. Tokens like stones were often used for keeping track of simple arithmetic. Even today, "counters" are widely used to record the score in a game, and schoolchildren in many countries still use a version of the abacus. Over time a "calculus" came to mean anything that was an aid to mathematics, whether it was a machine or just a handy technique. Since the 1700s, though, only one thing has deserved this name: *the* calculus, devised by Newton and Leibniz and developed ever since. It's fair to say that this tool, or set of tools, has touched every area of mathematics and almost every discipline that uses it. The ancient Romans called conjurors *calcularii*, for they too used pebbles, and perhaps the mathematical calculus is as wonderful as a magic trick — it's almost as impenetrable to outsiders, too.

The calculus comes in two flavours: differential and integral. Differential calculus gives us a way to describe the rate of change of something from its behavior: for example, if we know how the position of a car changes over time, we can use that to find how fast it was going at any particular moment. Integral calculus gives us new ways to add things up, enabling us to find areas and volumes that would be difficult or impossible otherwise. Both flavours work using limits (see Zeno's Dichotomy, page 18), and the Fundamental Theorem of Calculus tells us that, even though they seem to be about quite different things, they're intimately connected. In a sense the operations of differentiation and integration undo each other, just like the operations of multiplying and dividing.

Why Does It Matter?

The importance of differential calculus is the easiest to appreciate. It can convert a movement in space to the velocity of that movement; if the velocity changes it can find the acceleration, too. This isn't the place to describe how to actually do these calculations, but a concrete example might help. If you haven't met calculus before you'll have to take some of the steps on trust.

Suppose we drop a ball from a tower 200 m

26

(or 200 yd) high and make a video recording of it falling. We then analyze the video and determine that the height (*h*) of the ball above the ground after *t* seconds is approximated by

$$h = 200 - 4.9t^2$$

If we differentiate this once, we get a formula for the rate of change of its position at each moment, which is to say the velocity:

$$\frac{dh}{dt} = -9.8t$$

We got this answer using a few simple rules: the important thing is that the "fraction" *dh/dt* roughly means "the amount the height is changing when

time changes just a tiny bit," which in common-sense terms is "how fast the ball is falling at that moment." Notice it depends on time *t*, because the ball is speeding up as it falls. It's negative because it's going downwards, so the height is decreasing as time increases.

We can do the same thing again, to find the "rate of change of the rate of change of position," which sensible people call "acceleration":

$$\frac{d^2h}{dt^2} = -9.8$$

The notation on the left is a bit weird-looking; there's no significance to it except that we're now looking at the "second derivative." Notice that the right side no longer depends on time; the acceleration of the ball is a constant. That's because it's just the acceleration of the ball due to gravity, the only force acting on it as it falls (see Universal Gravitation, page 60, and Newton's Second Law, page 56,) and gravity isn't changing. This agrees with Galileo's legendary experiment: if you drop two balls of different weights from the Leaning Tower of Pisa, the heavy one lands at the same time as the light one (see illustration on page 28). Gravity is acting on both in exactly the same way, and weight doesn't enter into it.

We can even make sense of third derivatives in this simple physical setup. Generally, the rate of change of acceleration is called "jerk," which correctly suggests that it's an unpleasant thing to experience. In the case of our falling ball, at least until it hits the ground, it turns out that the jerk is zero. You can see this yourself: the acceleration is constant (it doesn't depend on the time variable *t*). This means that it doesn't change, so its rate of change must be zero!

This is all well and good, but very often we have the reverse problem: we know the acceleration of something, for instance, and we want to know what it does (that is, where it goes and how fast). This is where the Fundamental Theorem comes in, because it allows us to use the techniques of integration to "climb back up" from a rate of change to the thing that's changing. As you might

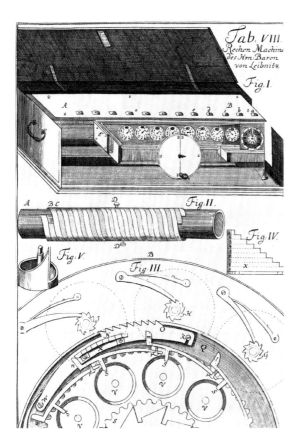

Leibniz's calculating machines could perform all four basic arithmetic functions.

The Fundamental Theorem of Calculus

imagine, there's a bit more to it than that, but that's the essential idea.

In More Detail

This "calculus" took a long time to emerge: some of the essential ideas had been sloshing around for thousands of years before 17th-century scientists finally pulled them together. One thing that made this possible was the arrival of algebra from the Arabic world and its combination with geometry by Descartes and his contemporaries (see Pythagoras's Theorem, page 10.) It took another hundred years to get the calculus into any kind of order, transforming it from a rather unruly bag of tools into a grand, unified system. That system may very well deserve to be called one of the great imaginative achievements of human culture.

The problems calculus solves are hard because they involve continuous variation, and it succeeds because it describes ways to approximate the solutions that can be refined and refined, so that these refinements approach a limit (see Zeno's Dichotomy, page 18) that is the precise answer we're after. This precise answer, in turn, can often be calculated right away using a few simple rules.

As an example, let's return to the falling ball of the previous section, except now imagine we know that the velocity of the ball at time t is given by

$$v = -9.8t$$

We know it started 200 m up and we'd like to know how long it takes to hit the ground. Let's approximate. We'll start at 200 m and calculate the velocity every, say, tenth of a second. Then we'll assume it was travelling at exactly that velocity for the whole tenth of a second until the next calculation; we'll work out the new position, then repeat. When $t = 0$, $v = 0$, so nothing happens (there's an instant when the ball just hangs in the air when you let go). When $t = 0.1$, $v = -0.98$ ms^{-1} (that's 0.98 meters per second, travelling downwards), so in the next tenth of a second we guess it will travel 0.098 m downwards,

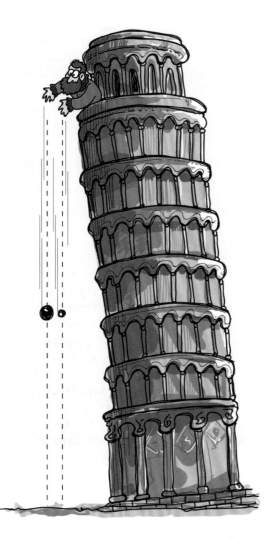

Galileo's experiment suggested, contrary to Aristotle, that all falling objects accelerate at the same rate regardless of their mass.

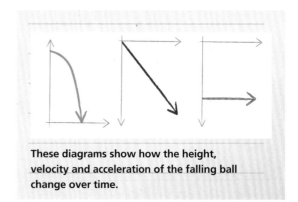

These diagrams show how the height, velocity and acceleration of the falling ball change over time.

The Shape of Space: Geometry and Number

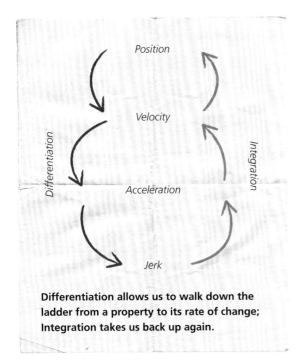

Differentiation allows us to walk down the ladder from a property to its rate of change; Integration takes us back up again.

velocity was constant for each of those, even though we know it wasn't. We could make our estimate more accurate by increasing the number of slices we carve the time up into; for example, we could use hundredths of a second instead of tenths:

$$h \approx 200 + \sum_{i=0}^{640} -0.098i$$

The essential idea of integration is that these are approximations of a correct value, which we get closer and closer to as we divide time into ever smaller slices; the true value must be the limit of this process when taken "to infinity." Then the sigma becomes a long "S" (still for "sum") and we get the integral

$$h(t) = 200 + \int_0^t -9.8t\,\mathrm{d}t$$

which gives us the height at a specified time t. Using some more simple rules we can express this as

$$h(t) = 200 - 4.9t^2$$

which is where we started. This ability to move from rates of change back to the thing that's changing is what the Fundamental Theorem of Calculus gives us. In a broader sense, though, it's much more than that, because it truly captures something fundamental about calculus itself. Its modern incarnation, Stokes's Theorem, applies to much more exotic situations than the ones we're considering. It says something rather profound about the natures of space and time themselves, at least as they appear in our mathematical models.

leaving it at a height of just about 199.902 m above the ground. We can continue this until we get a value below 0, at which point we know it's hit the ground. This takes 64 steps, so about 6.4 seconds (you can try this yourself using a spreadsheet if you like).

We can write down what we did like this, with the squiggly equals sign meaning "approximately equal to:"

$$h \approx 200 + \sum_{i=1}^{64} -0.98i$$

Now, look at it this way: we divided the total time of the fall into 64 parts, and pretended the

Differentiation enables us to say something precise about how things change; integration adds up apparently infinitesimal quantities. The miracle is they're really two sides of the same coin.

Curvature

Differential geometry lies at the heart of modern physics: curvature was one of its first and most fruitful concepts.

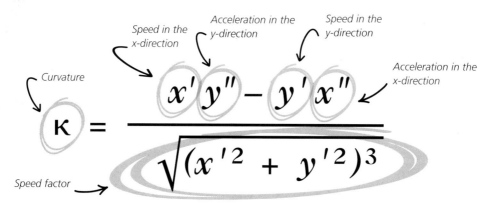

$$\kappa = \frac{x'y'' - y'x''}{\sqrt{(x'^2 + y'^2)^3}}$$

Curvature — κ

Speed in the x-direction — x'

Acceleration in the y-direction — y''

Speed in the y-direction — y'

Acceleration in the x-direction — x''

Speed factor — √ denominator

What's It About?

One way to imagine a curved line is as a set of fixed points that satisfy some equation or other (see Conic Sections, page 16). Another is more dynamic: imagine you're a tiny particle whizzing along the curved line at a steady speed and ask yourself what it would be like. Does it make a difference what shape the line is? Of course — compare driving on a long, straight desert highway and driving on a twisting country lane. Even with your eyes closed, and no idea where you are, you can feel the difference. A lot of what you're feeling is the effect of curvature, which gives us a precise measurement of, well, how curvy the road is at each point. Going around a bend forces us to move in quite a different way from when we travel in a straight line.

Why Does It Matter?

When you go around a curve, you're changing direction. That means you're being accelerated, and that in turn means you're experiencing a force acting on you (see Newton's Second Law, page 56). Cruising down a straight road, you may even forget you're moving at all, while that winding lane throws you vigorously from side to side. The curvature formula unites velocities and

accelerations in a delicate balance that precisely captures this curviness.

Driving on a long, straight road is comfortable but a little boring. It would make a terrible rollercoaster. Designers of funfair rides use curvature calculations to create exciting but safe levels of acceleration as you travel along the track, which is especially important in those sections of the ride that "loop-the-loop." Early designs simply spliced a circle onto a straight section of track, but that meant you made a sudden transition from zero curvature (straight track) to the circle, and the result was "jerk." You can probably imagine what that felt like — not pleasant, especially if you'd just scarfed down a hot dog and slightly too much cotton candy. You sometimes get the same experience on old sections of railway, which also used parts of circles (this time flat on the ground, of course) when the train needed to change direction. The jolt as you hit one of these is pretty jarring.

Today this problem is well understood, and thanks to differential geometry we have fewer of these sudden transitions. Our trains and rollercoasters use special curves that allow for a rapid but smooth change of curvature: an example is the clothoid. Similar considerations can be used

to find flight paths for spacecraft and airplanes that are as efficient as possible without subjecting them to dangerous forces; this is especially a concern for jet fighters whose high-speed trajectories can put huge strains on both machine and pilot.

More sedately, curvature is also used to calculate the focal points of curved mirrors and lenses. For a thick lens with constant (but not necessarily equal) curvatures on front and back, the "lensmaker's equation" gives us

$$P = (n-1) \left(\kappa_1 - \kappa_2 + \frac{(n-1)\, d\, \kappa_1\, \kappa_2}{n} \right)$$

where P is the power of the lens, n is the refractive index of the material, d is the thickness of the lens and the two curvatures are for the surfaces nearest and furthest from the light source, respectively. Notice that the strength of a lens depends only on what it's made of, how thick it is and its curvature.

In More Detail

The position of a point on a flat surface can be expressed by a pair of numbers called its coordinates (see Pythagoras's Theorem, page 10). If the point is moving, those x- and y-coordinates are changing over time, so it makes sense to talk about the rates of change in the x- and y-directions. Using

A sudden change in curvature, as when a straight track joins a circular one, causes jerk. Modern rollercoasters use curves called clothoids to smooth out the transition.

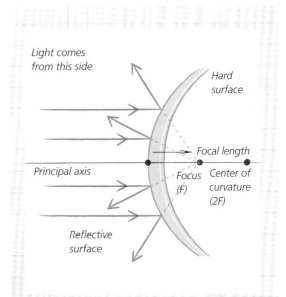

Light comes from this side

Hard surface

Focal length

Principal axis

Focus (F)

Center of curvature (2F)

Reflective surface

The curvature of a mirror determines its focal point. In this case, light is scattered by the convex surface, creating a distorted image.

the language of calculus (see The Fundamental Theorem of Calculus, page 26) we can write these in shorthand as

$$x' = \frac{dx}{dt}$$

and

$$y' = \frac{dy}{dt}$$

The little dash at the top — pronounced "prime," weirdly enough – means it's the first derivative with respect to time, which is to say the speed, in that direction. A second prime means acceleration (that is, second derivative), so x'' means the acceleration in the x-direction. For a given instant in time, we can calculate x and y (the position of the particle), x' and y' (how fast it's moving in the x- and y-directions) and x'' and y'' (the two components of acceleration). This information allows us to calculate the curvature of the path at that moment.

Looking at the equation on page 30, then, the top part of the fraction says "Multiply speed in the x-direction with acceleration in the y-direction, then do the opposite and find the difference between them." This gets very close to capturing curvature. The bottom of the fraction is largely there to adjust the equation, so it comes out the same whether the point is moving fast or slowly: after all, how curvy a road is has nothing to do with how fast you drive along it, although your subjective experience of the curves will of course be different at different speeds. The bottom part "factors out" that difference and leaves us with an objective measure of curvature.

We can visualize the curvature in a very simple way. At each point along the path there's a unique circle that nestles perfectly into the curve. This is its "osculating circle" (osculare is the Latin verb for "to kiss"). We can find it very easily: draw a line sticking straight out of the curve whose length is $1/\kappa$, and that will be the radius of the osculating circle. Notice that the more curved the path is at any point, the smaller the circle will be, which makes sense. When taking a sharp bend, your car's turning circle had better be no bigger than the osculating circle at that point!

With a bit more work we can define various curvature measures for two-dimensional surfaces. The easiest is "mean curvature." To see how it works, imagine standing on some hilly terrain. Turn around through a full circle: you'll see that the slope of the ground varies depending on which way you look. For example, one moment you might be facing downhill, a little later you'll be looking

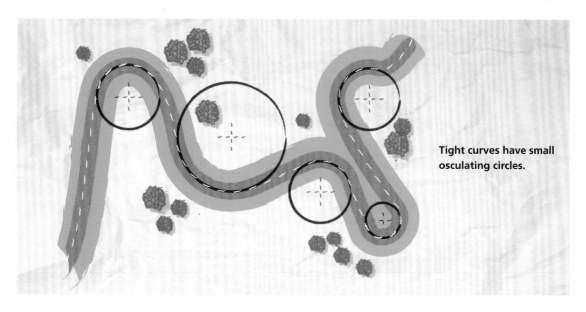

Tight curves have small osculating circles.

The Shape of Space: Geometry and Number

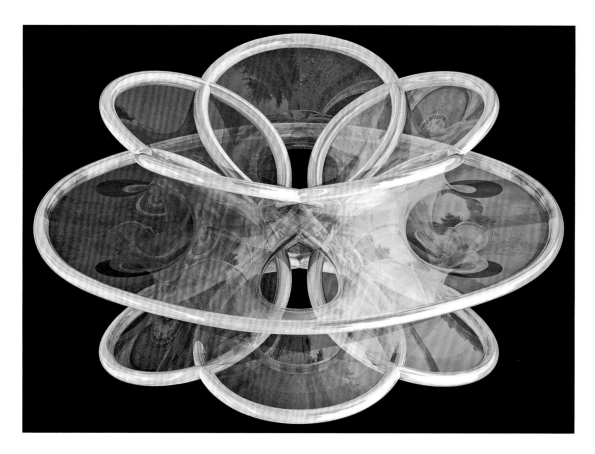

uphill. Similarly, imagine drawing a straight line out in front of you along the ground as you turn. You can calculate the curvature of this line, and it might well vary as you turn to look at different parts of the terrain. Mean curvature is just the average of the maximum and minimum curvatures you can find at that point.

Some natural processes tend to produce shapes with zero mean curvature called "minimal surfaces." The most familiar examples are soap bubbles and films, which tend to snap into minimal surfaces because of the surface tension within the film.

The surface tension in a soap film causes it to "find" the shape that minimizes its surface area. It turns out that this also makes the total mean curvature vanish.

The same effect gives similar shapes to tents and has been imitated by sculptors, architects and commercial designers.

The mathematics of minimal surfaces began with the maturing of calculus in the late 18th century, but it's still an active field of research with applications in many areas of science.

With calculus, geometry became far more expressive. It was able to give precise and useful expression to notions like curvature that had previously been vague and descriptive.

Frenet-Serret Frames

The curved path of a fly leads us to the trajectories of space probes.

How the tangent's changing ↝
$$\frac{dT}{ds} = \kappa N$$
Curvature, The normal

How the normal's changing ↝
$$\frac{dN}{ds} = \tau B - \kappa T$$
Torsion, The tangent, The binormal

How the binormal's changing ↝
$$\frac{dB}{ds} = -\tau N$$

What's It About?

Imagine you're a fly buzzing around in a room. Your head points in the direction you're going, and you can stick one of your right legs out to indicate which way you consider to be "right." You can also imagine wearing a conical hat that points in the direction that feels, to you, like "up," even while you're dive-bombing towards the cake platter.

That's three "directions" in space that make consistent sense to you, although to an outside observer they're constantly changing as you fly around. You can think of the directions as three arrows sticking out of you, all at right angles to each other. Technically the "forward" arrow is called the "tangent vector" (*T*), the "up" arrow is the "normal vector" (*N*) and the "left" arrow is the "binormal vector" (*B*). Together they form a "reference frame" that moves as you move and allows you to make sense of the space around you.

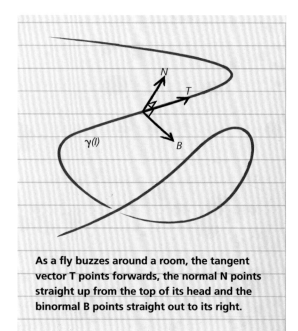

As a fly buzzes around a room, the tangent vector T points forwards, the normal N points straight up from the top of its head and the binormal B points straight out to its right.

The Shape of Space: Geometry and Number

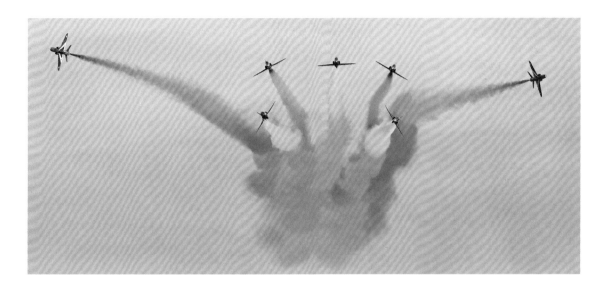

The calculation of the three vectors for a given moment in time depends on the curvature of the path κ (see Curvature, page 30) and τ, which represents "torsion," a measure of how much the curved path is twisting in 3D space. The formulas say that the motion at that moment depends only on those two things. Those fairly simple facts about the geometry of the path the fly is following tell you more or less everything you need to know about it.

In More Detail

The Earth is spinning at more than a thousand miles per hour, and it's going around the sun at a much greater speed than that. We can't find a spot in the universe that's not moving and make all our measurements from there. What we need instead is an "inertial frame," free of acceleration, and a mathematical language for translating from mine to yours and back just in case yours is different. As a result, reference frames — including ones that move — are almost as old as modern physics, dating back at least to Galileo. Choosing a convenient frame can make the difference between an easy problem and an intractable one.

The vapor trail left by an aircraft shows its curved path; the Frenet-Serret frame gives us a way to look at this curve from the pilot's point of view.

What makes Frenet-Serret frames special is that they give us a way to describe the geometry of curves, including the paths taken by moving objects, from the inside, as if the curve were a rollercoaster and the frame described what it feels like to ride it; as a result they often simplify calculations, too. Think of them as saying "if you're at a particular point in the movement and want to get to the next point, here's how to orient yourself to do that." Although "the next point" doesn't quite make sense as a concept, it's close enough for an intuitive picture.

Mostly we think about space being three-dimensional, and points moving around it in curves defined by time. We can, though, also think about moving around in Minkowski spacetime, which is four-dimensional; in that case the Frenet-Serret frame is more complicated and requires four vectors, one for each dimension (see $E = MC^2$, page 88).

Our orientation in space changes as we move about. Describing our motion in terms of a moving frame of reference feels intuitively "right" and it often makes our equations easier to solve, too.

Logarithms

Invented in the 1600s to help navigation at sea, today these mathematical gadgets have thousands of applications.

The number being logged

The number c that makes this true

$$\log_b (a) = \{c \mid b^c = a\}$$

The base

What's It About?

Imagine a machine that has a hopper on top and a chute at the bottom. You drop a number in the top and out of the chute drops another number. The internal workings of the machine multiply the number by itself three times (never mind how, exactly) so if we drop in the number 5, say, the machine calculates 5^3 to give us $5 \times 5 \times 5 = 125$, which is what drops out of the chute at the bottom.

Let's say you find the number 64 lying at the bottom of the chute — what number originally went in? The answer's found by gently carrying the 64 over to the "cube root" machine and dropping it into the hopper; out pops 4, because $\sqrt[3]{64} = 4$, which is another way of saying that $4^3 = 64$. This means we have a machine that undoes whatever our original machine did. In mathematical jargon, the cube root is the "inverse function" of raising to the power of 3.

Now suppose we have a third machine that does something slightly different. When you drop a number in the top, it raises the number 3 to that power. So if we put a 5 in, the number calculates 3^5 and what drops out of the chute is $3 \times 3 \times 3 \times 3 \times 3 = 243$. Again, suppose we found a number such as 19,683 at the bottom of

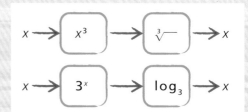

It's often helpful to think of a mathematical function as a box that takes in a number, does something to it, and gives us back a new number. Two boxes that cancel each other out, giving back the number we started with, represent inverse functions.

the chute: what number must have gone in? In other words, what number x makes $3^x = 19,683$ true? This is the question logarithms answer: $\log_3(x)$ is the inverse function of 3^x.

Why Does It Matter?

Logarithms were invented in the early 1600s to make it easier to multiply or divide large numbers by hand. This was particularly useful for navigators at sea, where a mistake could send them dangerously off-course. It turns out that logarithms

The third machine along raises 3 to the power of the number Betty drops in. To find out which number her colleague put in to get 19,683 out, she needs to put that number into the $\log_3$ machine.

turn problems about multiplication into problems about addition thanks to the way powers behave when you multiply them:

$$59,049 \times 2,187$$
$$= 3^{10} \times 3^{7}$$
$$= 3^{(10+7)}$$
$$= 3^{17} = 129,140,163$$

Using a book of log tables, mariners could reduce the problem above to the problem of adding 10 + 7, then convert the result back into the real answer; these tools work just like the machines with chutes and hoppers we were just imagining.

My example's artificially simple because the two numbers are exact powers of 3, but this technique turns out to work for any two numbers, and you can use any "base" instead of 3 to get the same result. For example, using base 5,

$$\log_5(59,049 \times 2,187)$$
$$= \log_5(59,049) + \log_5(2,187)$$
$$\approx 11.6043$$

$$5^{11.6043} \approx 129,140,163$$

The wavy lines indicate that these are approximate values, though they would (hopefully) be close enough for the purposes of the calculation.

Different bases are convenient for different uses. For example, the base 10 logarithm gives a way to count how many digits a number has when written in decimal notation — it will always be one more than the base-10 log of the number, rounded down. In the simplest case, we have numbers like 10,000, which is 10^4, so $\log_{10}(10,000)$ = 4, and 10,000 does indeed have 4 + 1 = 5 digits. More generally we have, say, $\log_{10}(37,652) =$ 4.5757881…, and rounding down and adding 1 again gives 5, the right number of digits. This number is often called the "order of magnitude," and for some phenomena it's more enlightening to look at the order of magnitude of a measurement than the measurement itself.

One case is the strength of an earthquake — the Richter scale is expressed in $\log_{10}$, meaning a magnitude 3 earthquake is 10 times more powerful than a magnitude 2 one. Other well-known examples that use $\log_{10}$ are the decibel scale of loudness and the pH scale for acidity. In all these examples, just calculating the numbers would make them hard to compare because you can get very big and very small values at different ends of the scale. Applying the $\log_{10}$ function makes them easier for human beings to understand, even those who don't know anything about logarithms.

Another frequently used base is 2, which counts how many times something has doubled. One hidden appearance of $\log_2$ is in musical octaves: a pitch sounds one octave higher than another if its frequency is double.

For historical reasons, the pitch called A4 (around the middle of the piano keyboard) is usually tuned to 440 Hz. Suppose we have a pitch of 110 Hz; we have

$$\log_2(440) - \log_2(110) = 2$$

so there are exactly two octaves between them (110 has to be doubled twice to get to 440).

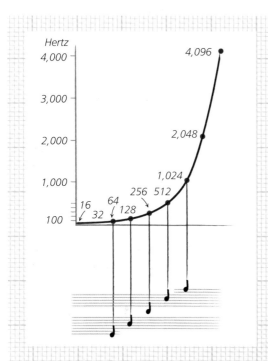

Musical notes separated by an octave form a sequence of powers of 2, so logarithms can be used to count how many octaves separate a pair of pitches.

Base-2 logarithms are also used to calculate the half-lives of radioactive materials, where again growth (or this time, decay) by a factor of 2 is the key ingredient.

In psychology, Fechner's Law states that the intensity of our perceptions tends to vary by the natural log of the intensity of the stimulus. It's a rather general, approximate sort of claim but it works well for a surprisingly wide range of types of stimuli. Perhaps it explains why we prefer to measure many things we experience through our senses on logarithmic scales.

In More Detail

Very often a special base is used, known as e, the so-called "base of natural logarithms." This isn't a simple number like 2 or 10 but an "irrational" number that can be thought of as an infinitely long decimal that never repeats itself. Roughly speaking, it arises from thinking about the kind of continuous growth we find in much of the natural world.

A common way to derive it is by thinking about compound interest. First, imagine you've put $1,000 in a savings account that pays 5% interest per year. How much will be in the account after 3 years? It turns out the calculation is:

$$1,000 \times (1 + 0.05)^3 \approx 1,157.625$$

That is, you take your initial sum of money and apply the interest three times in succession. Well, that assumes the interest is applied once per year — is it better to get half the interest, twice a year? You bet it is!

$$1,000 \times (1 + 0.025)^6 \approx 1,159.69$$

Here we've taken half the interest rate (2.5%, which is 0.025) and applied it in 6 batches instead of 3 batches. You get a little more money this way. Could we keep doing this and get more and more money? No: the amount of interest will instead approach a limit.

Natural processes don't grow in instantaneous jumps: they're continuous. The ancient slogan for this was *natura non facit saltum*: nature does not

The Shape of Space: Geometry and Number

make leaps. Even apparently sudden events seem, when you zoom in on them enough, to be quick but smooth transitions rather than instantaneous changes. This is one of the basic assumptions behind Newtonian science. Logarithms take us from a simplistic picture involving a sequence of sudden, discrete changes to something more like a real, organic process.

To see what that would look like, imagine the number of interest payments in the year steadily increasing, making them more and more crowded together, and the amount of interest each time getting smaller. Perhaps you get interest every week, every day, once an hour, even once a second … we keep on making the payments more and more frequent while decreasing the amount each time.

This idea leads us to consider the limit (see Zeno's Dichotomy, page 18):

$$\lim_{n \to \infty} \left(1 + \frac{1}{n}\right)^n$$

Here, the number of increases in the year approaches infinity while the size of the increase each time approaches zero. This limit is the number we call e; it describes continuous growth of the kind we seem to see in the natural world. Its value is about 2.718, but can never be expressed perfectly as a decimal or a fraction. That's the base we use when we want to express this type of growth.

Let's see a very quick example. Suppose I have a plant that was 40 cm tall 7 days ago, and is now 45 cm tall. What's its growth rate? Well, let's assume it grew continuously; then its growth rate over the week was $\log_e(45/40) = 0.1178$ (approximately) or 11.78%.

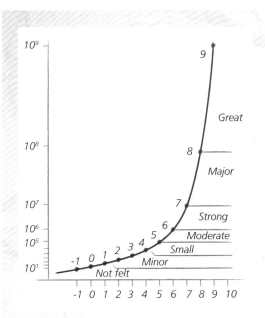

The Richter scale is logarithmic, which captures our experience of increasing intensity much better than a linear scale would.

Note that it's actually grown by 12.5%; that's the growth rate you'd need if it jumped up by 5 cm all at once at the end of the week. But it didn't do that. Instead it grew continuously at a rate of 11.78%. If this continues then in another week it'll be this tall:

$$45 \text{ cm} \times e^{0.1178} \approx 50.625 \text{ cm}$$

This ability to make exact predictions about continuous growth and decay processes makes the weirdness of natural logarithms worth getting used to.

Being able to undo the act of raising a base to a power is helpful, but logarithms give us much more: a model of continuous growth and a neater way to represent exponentially growing phenomena.

Euler's Identity

**This intimate connection between five fundamental numbers
is often called the most beautiful equation in math.**

The square root of –1

The ratio of a circle's
circumference to its diameter

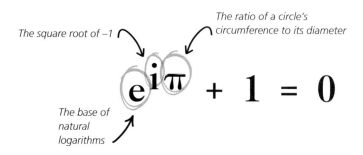

$$e^{i\pi} + 1 = 0$$

The base of
natural
logarithms

What's It About?

This equation is a simple and surprising relationship between five basic numbers that ought to be completely unrelated to each other: e, the base of natural logarithms (see Logarithms, page 36), i, the "imaginary" square root of –1 (see below), π, the ratio of a circle's circumference to its diameter, and 1 and 0, the numbers that leave other numbers unchanged when you multiply by them and add them, respectively.

What an amazing coincidence that these fundamental numbers, which seem to have nothing to do with each other, should come together in this way! Of course it's not really a coincidence, but the elegance of the equation and the surprise it produces make it one of the most famous in the world.

Why Does It Matter?

Euler's Identity is about the so-called "complex numbers," a number system that goes beyond the ones we're all familiar with from our schooldays (see the next section). Complex numbers take a bit of getting used to and they behave rather weirdly. In fact, for a long time they were thought to be

mere curiosities, the whimsical inventions of out-of-control mathematical imaginations.

It turns out, though, that complex numbers are the building blocks of a huge amount of physics and engineering. One reason is that, under certain conditions, we can always solve our equations. Even if there's no "real" solution, we're guaranteed to have the expected number of "complex" ones. This, on its own, streamlines many things that would otherwise be very complicated, and that in turn makes it possible to build more elegant theories.

Many practical problems are much easier to solve using complex numbers (and Euler's Identity) than they would be using ordinary, everyday numbers. They make subjects like fluid dynamics (see The Navier-Stokes Equation, page 96), electronic engineering (see Maxwell's Equations, page 92) and digital processing (see The Fourier Transform, page 138) work smoothly; they're also found in the fundamental equations of quantum mechanics (see The Schrödinger Wave Equation, page 104). In many cases these applications have their origins in differential equations (see Newton's Second Law, page 56), which are found everywhere in modern science and technology and which are much easier to work with in the world of complex numbers.

In More Detail

When you multiply a number by itself, the answer's always positive. This is true even if the number's negative: so −2 × −2 is 4, not −4. Sometimes this surprises people and if you're one of them you should know that all the way into the 1700s mathematicians still argued about which was correct, and some suspected that negative numbers were nonsensical.

Eventually, though, it was realized that accepting this rule made everything else work out OK. A consequence: there are no square roots of negative numbers, because no number times itself gives a negative answer (see Pythagoras's Theorem, page 10). Without digging into the history, we might simply ask this: what if it *did* make sense to write things like $\sqrt{(-4)}$? After all, someone once asked what happened if we pretended — just for a moment — that we could take 5 away from 3, though history doesn't record their name.

It's enough to use the letter i as shorthand for

$\sqrt{(-1)}$, since some standard rules of algebra enable us to write other negative square roots in terms of it:

$$\sqrt{(-64)} = \sqrt{(64 \times -1)}$$
$$= \sqrt{64} \times \sqrt{(-1)} = 8i$$

The letter i stands for "imaginary," because nobody thinks these are proper numbers; or at least, they didn't at first. Over time the system of numbers they give rise to, called the "complex numbers," turned out to be very useful in practical situations as well as in pure math, and gained widespread acceptance. We meet them a couple more times in this book and although we never need to look at them in any detail it doesn't hurt to have a rough idea of how they work.

Euler's Identity is a basic fact about the geometry of these numbers. Essentially, it says that rotating the number 1 through half a circle gives you −1, but this must seem like a very cryptic thing to say. First, you can think of a complex number as being of the form $a + ib$, where a and b are just ordinary numbers (including negative numbers, fractions and so on). This is why these numbers are called

It's unusual to see five such different numbers getting along so well together.

Euler's Identity

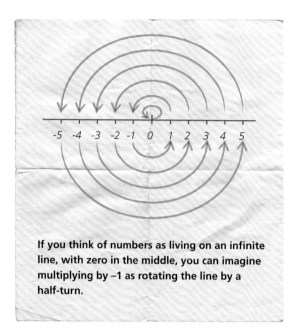

If you think of numbers as living on an infinite line, with zero in the middle, you can imagine multiplying by –1 as rotating the line by a half-turn.

axes cross) to that point — the line you'd follow if you walked straight from the origin to the point in question. Suppose you start off standing on the origin and looking along the x-axis in the positive direction: one way to describe a number is the angle you need to turn to face it, and then how far you need to walk in a straight line to reach it. These two numbers are called the "argument" and "modulus," respectively, and they give you another way to represent the complex number.

It turns out that we can use this information along with the special number **e** to represent complex numbers in a particularly compact way. You'll have to trust me on this, but if a complex number has argument a and modulus m we can express it as

$$m\,\mathrm{e}^{ia}$$

Now, imagine keeping m fixed and letting a increase; what happens is that the line representing our complex number begins to turn like the hand of a clock, tracing out a circle whose radius is m. This relationship with circular motion is what makes complex numbers pop up in all kinds of situations (see Trigonometry, page 14).

"complex," which means "made of more than one part" rather than "complicated." The a part is called the "real part" of the number, and the b is the "imaginary part."

If you want to represent an ordinary number using this system, you can just set the value of b to zero, so that there's no imaginary part. Then you can put your number, which is just a, on a number line that looks like an infinitely long ruler with zero in the middle, the positive numbers stretching off to the right and the negative ones to the left.

Complex numbers with an imaginary part don't fit on a number line, but they can be mapped out on a two-dimensional plane in something called an "Argand diagram." Every point on the plane has x- and y-coordinates (see Pythagoras's Theorem, page 10), and we can interpret these as the real and imaginary parts of the complex number. This means that the diagram contains a copy of the ordinary number line (the x-axis), but a lot more numbers, too.

To make sense of Euler's Identity, though, we need to look at the Argand diagram slightly differently. Just as every complex number can be thought of as a point on the plane, it can also be thought of as a *line* from the origin (where the

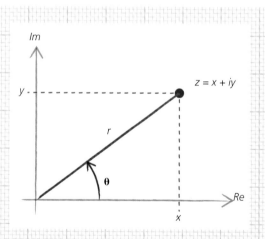

A complex number can be thought of as a point in 2D space. We can describe how to get to it from the origin using an angle (θ) and a distance (r).

The Shape of Space: Geometry and Number

Most fractal images are Argand diagrams and Euler's identity often makes them much easier to calculate.

We need one more idea before Euler's Identity will make sense. You're probably used to seeing angles measured in degrees, with 360 of them in a full turn. This, as you may have noticed, is ridiculous. Why 360, exactly? Allegedly it has something to do with the Babylonians, but that's never been much of an excuse for anything, and degrees really aren't very convenient or enlightening a lot of the time. The alternative, which is used very widely in mathematics, is to measure angles in radians. In this system one full turn isn't a whole number of radians, it's 2π of them. Though this seems strange at first, it intimately relates angles to circles, which is how things ought to be.

So, what number does this represent?

$$e^{i\pi}$$

Since we don't show any number in front we can deduce that the modulus is 1. Since there are 2π radians in a full circle, we can surmise that π radians make half a circle. So this is the complex number that's 1 unit long and turned through a half-turn. That number is –1. So,

$$e^{i\pi} = -1$$

from which it surely follows, as night follows day, that

$$e^{i\pi} + 1 = 0$$

That's Euler's Identity: an expression of the central role of circular motion in the world of complex numbers.

It's not the most obviously useful equation in this book, but Euler's Identity unifies many different subjects (trigonometry, complex numbers, logarithms). It's also handy when calculating with complex numbers.

Euler's Identity

The Euler Characteristic

The Four-Color Theorem can be understood by anyone, but is still one of the hardest mathematical results of all time.

The Euler characteristic → χ = *Vertices* → V − *Edges* → E + *Faces* → F

What's It About?

You've probably seen the famous "utilities problem": given three houses and three utilities — water, gas and electricity, say — the puzzle is to connect each house to each utility without crossing your lines. It's a common puzzle to give to children, which is pretty mean because it can't be solved. The trouble arises from a quantity called the Euler Characteristic which, on a flat surface like the one the puzzle's usually presented on, is 2. A solution *is*

possible on a donut (that is, a "torus"), where the Euler Characteristic is different.

Perhaps you've also heard of another problem: if I give you any map (including an imaginary one) that divides up its flat area into "countries," what's the smallest number of colors you can use to colour it in, making sure that no two countries that share a border are the same color? This is the so-called "map-coloring problem," and it, too, has a close relationship with the Euler Characteristic.

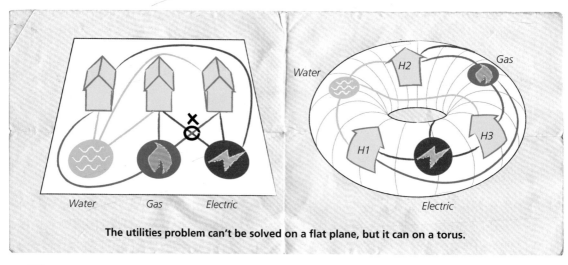

The utilities problem can't be solved on a flat plane, but it can on a torus.

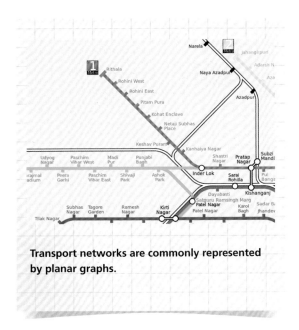

Transport networks are commonly represented by planar graphs.

into several regions, bounded by its edges; each of these is called a "face." Clearly, there's some relationship between the number of vertices, edges and faces a planar graph can have; try drawing some out and see for yourself. The reason the utilities problem can't be solved is that the graph you'd need to draw to solve the puzzle is non-planar, so you can't actually draw it on a flat sheet of paper.

We can calculate the Euler Characteristic of a plane by drawing a graph on a page, with no crossing edges. If we draw a triangle, for instance, we use three vertices and three edges and divide the plane into two faces, the inside and the outside. So we have

$$\varkappa = 3 - 3 + 2 = 2$$

The Euler Characteristic of the plane is 2. If you're thinking there's something special about triangles you'd normally be right, but in this case there isn't. Try the same calculation with other graphs and you'll get the same answer, and perhaps get an insight into why it's so.

Suppose you have a complicated map to color in. How many colors will you need? This question is certainly easy enough to understand, and not too hard to translate into a problem about graphs: just put a vertex inside each country and connect two vertices if the countries share a border. That we never need more than five colors for any map was proved way back in 1890 by British mathematician Percy Heawood, but it was always suspected that four would do the trick. The proof finally came in 1976, but to produce it Kenneth Appel and Wolfgang Haken of the University of Illinois had to do a lot of number-crunching by computer, a method that remains somewhat controversial since no human would ever be able to check it.

In More Detail

The utilities problem is about what mathematicians refer to as a "graph," which is a system of points called "vertices" and lines joining them called "edges." Graphs are very useful; they can represent any network of little things (the vertices) connected by paths, cables or tubes.

Public transport maps are, essentially, graphs, and so are diagrams of computer networks, industrial processes, circuit boards and so on — so graphs have many practical applications. In many important cases we have only very slow, effectively brute-force methods to solve them — a big discovery in this area could have a significant impact on our lives.

A graph is called "planar" if it can be drawn on a piece of paper (which is flat, like a mathematical plane) without any of its edges crossing each other. When this is done, the graph cuts the page

---

Topology studies the properties of spaces, from how they're connected to the kinds of holes they have. The Euler Characteristic turns these problems from mere descriptions into equations.

The Hairy Ball Theorem

An abstruse fact about vector fields on topological manifolds tells us why there's always a place on the Earth where the wind isn't blowing.

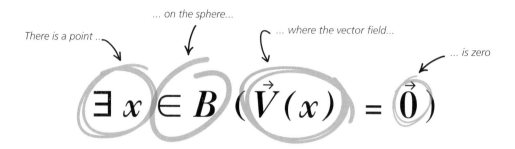

There is a point ... *... on the sphere...* *... where the vector field...* *... is zero*

$$\exists\, x \in B \;(\vec{V}(x) = \vec{0})$$

What's It About?

The Hairy Ball Theorem (HBT, for short) says that there's always a place on the surface of the Earth where the wind's not blowing. We're used to seeing the wind speed and direction at a point on a map represented by a little arrow attached to that point, with the length of the arrow representing the speed. Usually we only look at a small area of the globe, flattened out into a map (see The Mercator Projection, page 110). But if we look at the wind in this way on the whole globe, the theorem tells us there'll always be a spot, somewhere, where the arrow vanishes because the wind just isn't blowing there.

You can't even come up with a way the wind could be blowing, in theory, that doesn't have a spot like this. If you manage to get rid of it and make the wind blow there, it'll just pop up somewhere else in the world. This has nothing to do with how weather systems behave: it turns out to be a basic geometrical fact.

Why Does It Matter?

Consider a spherical cat covered entirely in fur; the HBT says we can never comb the cat without creating a tuft somewhere. This doesn't sound

No matter how you brush it, a perfectly spherical cat covered entirely in fur will always have a tuft sticking up somewhere. Admittedly, this may seem like a rather academic point.

The Shape of Space: Geometry and Number

The flow of air over the Earth's surface is more like the hairs on a cat than might first appear.

immediately useful, and as it was a branch of differential topology the HBT was long considered a result belonging to the most rarefied heights of "pure mathematics," on which more practical things could depend only very indirectly.

The HBT has, though, found a number of applications. The most tangible relate to the case we've already described: the theorem puts a limit on the possible ways in which air — or some other fluid, such as water — can flow continuously over a surface. It also tells us that if you take a ball and rotate it, even if you turn it around in very complicated ways, there'll always be one spot that's exactly where it was when you started.

In physics, the HBT is important in the study of spherical waves such as light and sound, and in the study of electrical and magnetic fields. Direct technological applications of such a deep result as the HBT tend to come up at the cutting edge of research. For example, in 2007 technologist Gretchen DeVries and her colleagues at MIT were able to bond gold nanoparticles using the HBT, giving us a way to build up larger nanostructures, something like crystals or polymers; in 2010, Mark Laver and Edward Forgan published a paper in the British journal *Nature Communications* on the HBT's

effects on the behavior of superconductors. Both projects illustrate how pioneering technological ideas can arise from an equation that was once considered hopelessly abstract.

In More Detail

The HBT was first proved by the Dutch mathematician and philosopher L.E.J. Brouwer in 1912; it had been stated and, it is said, already given a proof by the French polymath Henri Poincaré shortly before. This was a time of enormous progress in topology, a new and seemingly very exotic domain of mathematics that Poincaré had been instrumental in inventing. Yet much of Poincaré's work was related to problems in physics; his work with the Dutch physicist Hendrik Lorentz contributed significantly to the development of Special Relativity. Brouwer was a rather less pragmatic figure, as much a philosopher — even, at times, a mystic — as a mathematician, although his contributions to math are significant.

So, leaving aside the spherical cats and so on,

what's the HBT really about? It's a fact about continuous tangent vector fields on a topological 2-sphere. Let's unpack these technical terms and see how they fit together.

A vector can be thought of as a little arrow; its important characteristics are how long it is and which way it's pointing. The zero vector is just an "arrow with zero length," if you can imagine such a thing. It's what you get when a vector gets shorter and shorter and finally disappears. The HBT is going to say that under certain circumstances we'll always be able to find at least one zero vector somewhere; making sense of those circumstances requires a bit more decoding of the jargon.

A vector field on a surface attaches an arrow to every point. These arrows are tightly packed without any gaps. Physicists use vector fields to model many phenomena, including electromagnetic and gravitational fields, as well as the flow of fluids such as air and water. At this point a good picture to have of a vector field is the arrows indicating wind direction and speed on a weather map. We just have to imagine there's an arrow at every point, not just some of them.

A tangent vector is one that's flattened down so that it points along parallel to the ground rather than sticking up out of it or down into it. Imagine a tiny Dalek wandering around on the surface of the Earth; the unfortunate vector at the point where it's standing tells it which way to point its plunger. As it

moves about, the arrows it's standing on may point in new directions, causing it to swivel around on its terrifying wheels.

A vector field is continuous if the arrows never change in size or direction by a sudden jump; even if they seem to, if we "zoom in" closer we should see that a quick, but smooth, change is going on. In most physical situations we like to assume that "nature does not make leaps." Quick change is fine; it's instantaneous change that we don't see often in the natural world. The HBT is only true if the vector field changes continuously as you move around on the surface of the ball.

So much for continuous tangent vector fields; what about topological 2-spheres? You can picture one by thinking of something like a balloon. Now imagine it's made of a very flexible rubbery material that can be stretched, squeezed, twisted and otherwise manipulated. As long as we don't burst the balloon it remains a topological sphere, even if it doesn't look very spherical any more. Topology studies the properties of shapes that stay the same even when we do these rather radical things to them. The HBT isn't true on other kinds of surface; it's not hard to comb a hairy donut flat, for example, although it's not very clear why you'd want to.

The "2" in "2-sphere" refers to the fact that this is a two-dimensional space, just as the surface of the Earth is. You might object that the Earth is

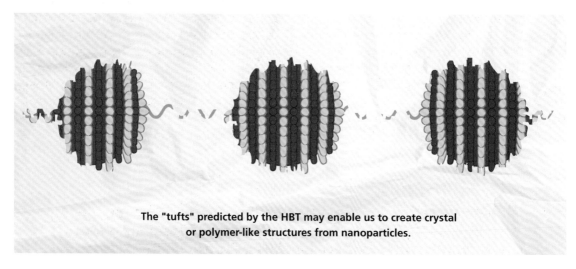

The "tufts" predicted by the HBT may enable us to create crystal or polymer-like structures from nanoparticles.

The Shape of Space: Geometry and Number

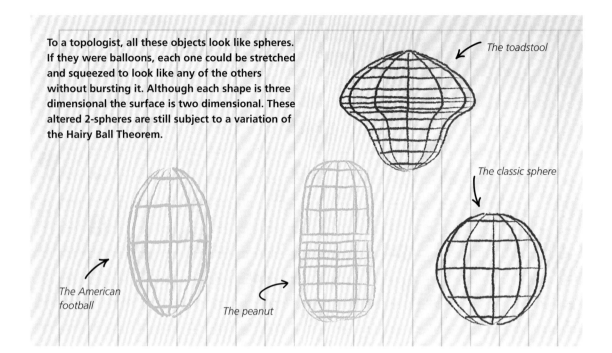

To a topologist, all these objects look like spheres. If they were balloons, each one could be stretched and squeezed to look like any of the others without bursting it. Although each shape is three dimensional the surface is two dimensional. These altered 2-spheres are still subject to a variation of the Hairy Ball Theorem.

The toadstool

The classic sphere

The American football

The peanut

a three-dimensional shape, and indeed it is, but its surface isn't; we can find our way around it using two dimensions such as latitude and longitude. Roughly speaking, this is what we mean by "dimension" — the more coordinates we need to identify a point in a space, the more dimensions the space has (see Pythagoras's Theorem, page 10). The HBT is a statement about surfaces, which have two dimensions, not one-dimensional lines or three-dimensional volumes.

Topologists use the word "sphere" in a rather technical way that allows it to have any (finite) number of dimensions, and the HBT isn't always true when the dimension changes. The "topological 1-sphere," for example, is a common-or-garden variety circle, and it's easy to come up with a vector

field on the circle that doesn't have any zero vectors: just have a little arrow sticking out at a tangent to the circle at every point.

Something called the Poincaré-Hopf Index Theorem, though, tells us that a version of the HBT is true for all spheres of even dimension; the next one would be the four-dimensional "sphere," but that's a tough thing to picture. As for the odd-dimensional spheres, including the circle, it turns out they can all have vector fields that don't disappear to zero anywhere. This kind of high-dimensional topology may seem to belong in the ivory towers of academia, but it's cropped up in recent work on interpreting huge sets of data, an increasingly pressing problem in the scientific and commercial worlds.

Vector fields are ubiquitous in modern physics. The topology of the space they're in determines fundamental things about which fields can possibly exist and which can't.

The Hairy Ball Theorem

A Mirror

up to Nature

SCIENCE

Kepler's First Law

Why do the planets move in ellipses rather than circles?

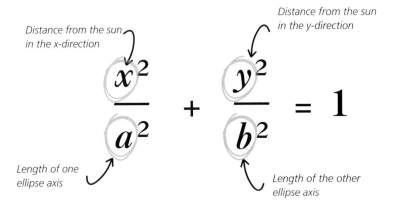

Distance from the sun in the x-direction

Distance from the sun in the y-direction

$$\frac{x^2}{a^2} + \frac{y^2}{b^2} = 1$$

Length of one ellipse axis

Length of the other ellipse axis

What's It About?

You can draw a circle like this: put a nail in the spot where the center should be, loop a piece of string around it, stretch it tight and draw wherever you can reach with the end of the string (this is also the principle of a compass). The curve of the circle gives all the places that are exactly the distance from the nail given by the taut string.

An ellipse looks like a stretched-out circle and, several centuries before Johannes Kepler came up with his law in 1605, artists had already discovered they could use ellipses to paint circles in perspective. You can draw one with a similar setup, but this time instead of one nail you want two, side by side, and the string looped over both of them — instead of a single center, each of these nails is called a "focus"

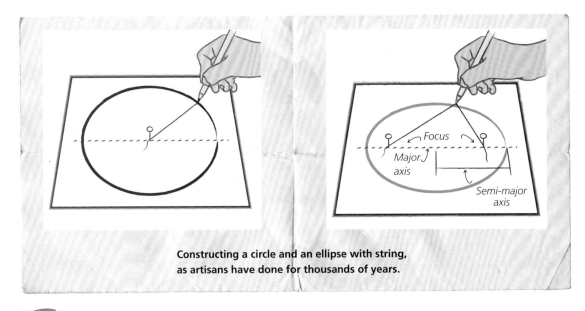

Focus

Major axis

Semi-major axis

Constructing a circle and an ellipse with string, as artisans have done for thousands of years.

A Mirror up to Nature: Science

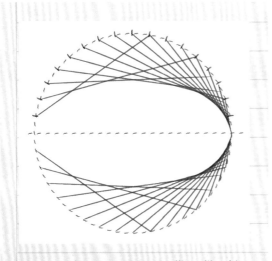

You can make an approximate ellipse like this, which is called a "thread construction." Each straight line is a path the planet might follow if the sun suddenly vanished.

of the ellipse. It's quite tricky to keep the string tight but this method has been used for centuries by carpenters, gardeners, builders and (more recently) the makers of crop circles.

Thanks to Kepler, we now know that the planets move around the sun in ellipses rather than circles; the sun is always at one focus. This law brought together the ancient geometry of conic sections (see page 16) with the new technology of telescopes to create the most precise model of the solar system in history.

Why Does It Matter?

Astronomers have made mathematical models of the universe since at least Ptolemy in the first century CE. His model had the moon, sun, planets and stars all moving around the Earth on a set of nested spherical shells. This would make their paths perfect circles, which led to predictions that disagreed with what was observed in the sky, so over time a set of increasingly complex adjustments was introduced to make the model more accurate.

Although a few writers suggested alternatives, this general approach was the mainstream view until the 1540s, when Copernicus proposed a model with the sun at the center and the planets rotating around it; the moon went around the Earth and the Earth also rotated on its axis. This concept is now very familiar to us, and it pulled together a number of ideas that had been proposed before into a system that was much simpler than the complex set of checks and balances the Ptolemaic model had become. Its predictions were only about as good as the old system's, but it was considerably easier to use.

Copernicus's model was a radical change, but quite a few things from Ptolemy remained: one was the idea that the heavenly bodies all moved in circular paths. It wasn't for another half-century or so that Kepler, trying to make sense of his own observations, hit on the idea of having planets move in ellipses. He had no evidence or explanation for this, but it modelled what he was looking at better than circles did, so he went with it.

It has to be said that the old Copernican system coped with planets fairly well because their orbits are actually very close to being circles. What it couldn't cope with was phenomena like comets that seem to come and go periodically — they flash past us close by and at great speed, but don't return for a long time. That doesn't seem very much like circular motion.

It turns out that, at least in our solar system, comets have much more obviously elliptical orbits than planets. This new model eventually enabled astronomers to calculate the orbits of comets and their "periods," solving the ancient problem of predicting when they would reappear. This was first done in 1705 by the English astronomer Edmund Halley, who used Newton's techniques to show that his comet returned every 76 years; this led to a correct prediction of its next appearance, a very impressive confirmation of the power of the new astronomy.

Like most laws in physics, Kepler's are really approximations: in fact the planets of our solar system don't follow perfect elliptical orbits. The approximation, though, is very close. What's more, anything orbiting in a gravitational field obeys it, so it applies to artificial satellites and space probes as well as to moons and planets.

In particular, think of a satellite in orbit. The only thing affecting its motion and stopping it from flying off into space is the Earth's gravitational field. In a sense, then, the satellite is in free-fall; it just never hits the ground. Its motion, like that of comets and planets, will be elliptical, this time with the Earth at a focus.

Just how stretched-out an ellipse we choose depends on what we want out of the satellite. Some follow nearly circular paths but a few are placed in "highly elliptical orbits." Rather like comets going around the sun, these satellites swing close to a point on the Earth and stay near it for a period of time while they're turning around, before zipping off on the rest of their journey. This makes them good for taking specific kinds of measurements. For example, the European Space Agency's Cluster II project put four probes into a highly elliptical orbit, where they're currently making a detailed map of the Earth's magnetosphere and studying the effects of the solar wind.

In More Detail

Notice that the equation on page 52 is that of a conic section (see Conic Sections, page 16),

$$A x^2 + B x y + C y^2 + D x + E y + F = 0$$

with $A = 1/a^2$, $C = 1/b^2$, $F = -1$ and all the other capital letters equal to zero, which makes those terms vanish:

$$\frac{1}{a^2} x^2 + 0xy + \frac{1}{b^2} y^2 + 0x + 0y + (-1) = 0$$

In Kepler's time, conic sections were relatively new to Western mathematics and Kepler may have been attracted to them for mystical as well as scientific reasons: he was obsessed by the idea of a rationally ordered cosmos governed by the "music of the spheres." It wasn't until the time of Newton, nearly a century later, that we had a good explanation for why ellipses were a good choice.

Kepler's final account of the orbits of the planets is usually summarized in terms of three laws. The second says that the speed of the orbit varies according to where the planet is in its elliptical motion: near the pointy ends the planet "whips around" and moves more rapidly than it does along the flatter part of the ellipse (see Conservation of Angular Momentum, page 62). The third says that the larger the orbit, the longer it takes to get around it, so a year on Mars lasts longer than one on Earth (a year on Mars is about 687 Earth days).

Kepler stretched out the circular orbits into elliptical ones.

A Mirror up to Nature: Science

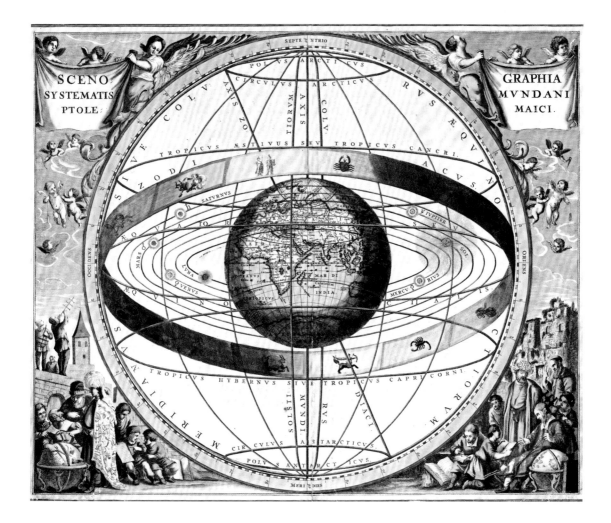

The Ptolemaic model of the solar system.

Those were rough statements, but they can be made mathematically precise. In that form all three can be derived from Newton's system of physics (see Newton's Second Law, page 56, and Universal Gravitation, page 60). Kepler's laws came first, based on observations and intuitions about how things ought to work; Newton only justified them much later. In modern presentations we tend to derive Kepler's laws from Newton's mechanics; that theory is in turn expressed in the language of vector calculus, a set of techniques Newton pioneered but that has taken on a life of its own since then. The derivation is still quite fiddly. It's remarkable that, without any of that, Kepler was able to make the imaginative leap to formulate these laws.

Conic sections aren't just mathematical curiosities. Not long after Copernicus, they provided Kepler with a beautiful model of the way the planets orbit the sun.

Kepler's First Law

Newton's Second Law

It's the second most famous equation in all of physics.

Force   *Mass*   *Acceleration*

$$F = m\,a$$

What's It About?

When I throw a ball on a windless day it moves in an arch-like curve: it goes up at first, reaches a high point and then starts to fall down again until it lands some distance away. If you watch someone throw a ball many times, it might strike you as odd that the ball, though it doesn't always follow the same path, always follows the same sort of path.

What mysterious natural law constrains it to do this, even if you throw it at different angles and with different amounts of effort?

Newton's Second Law is a statement about a simple relationship between force, mass and acceleration. It's still employed every day by engineers and scientists to calculate and make predictions; it's pretty easy to use and gives good

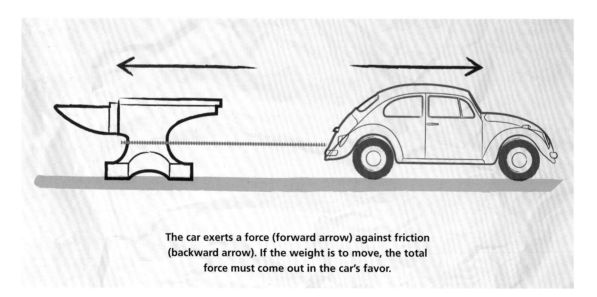

The car exerts a force (forward arrow) against friction (backward arrow). If the weight is to move, the total force must come out in the car's favor.

A Mirror up to Nature: Science

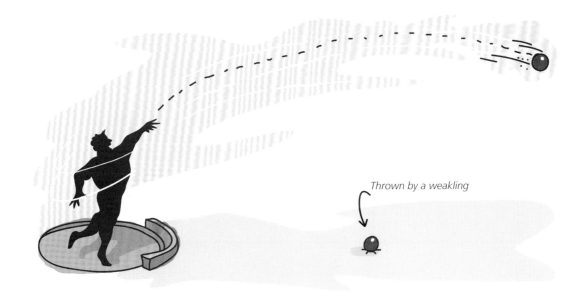

Thrown by a weakling

answers to a surprisingly wide range of questions about the physical world.

Many of the ideas in Newton's theory of motion were already around when he put them together. His achievement was to present the universe as a single mechanism in which this kind of regularity was expected, not puzzling, and had its roots in the way the whole system worked.

Why Does It Matter?

Newton's physics is one of the single most influential discoveries (or inventions, if you prefer) in the history of science. It unified observations and mathematical ideas from the preceding century into a single, coherent system that could be used to solve a wide range of problems. Despite Newton's famous interest in alchemy and astrology, his physics avoided anything murky and mystical: it was clean, precise and modern. This aspect of his work was a huge influence: it was a style of doing science that must have seemed like a breath of fresh air.

Newton's system is often summarized in terms of three laws, which is where this one gets its name. The first law says that if nothing's happening to an object, it moves at a constant velocity, meaning at a constant speed and in a straight-line direction. So, if you throw a ball in the void

The stronger you are, the more force you can put into the throw, and the more acceleration you'll give it.

of space it will just travel in a straight line forever, never speeding up or slowing down, unless some other force (such as gravity) starts to act on it.

Newton's third law is best-known by a famous slogan: "Every action has an equal and opposite reaction," which is more or less word for word what Newton's original Latin says. When you put a coffee cup on a table, gravity is pulling the cup down towards the floor. Naively, what's keeping the cup from falling is the equal and opposite reaction of the table on the cup. On the face of it, this looks a bit mysterious: how does the table know to push just the right amount to stop the cup falling downwards?

Suppose you're trying to push-start a car while wearing roller-skates. When you first push, the car seems to push you backwards, and actually that's a slightly spooky phenomenon. How does it do that? Well, most of us would say it's not the car pushing you but you pushing yourself. You might protest that you're pushing forwards, which surely can't lead to a movement backwards — what Newton's Third Law effectively says is that yes, it can, just as everyday experience attests.

Newton's Second Law

In More Detail

The easiest way for us to get a grasp of Newton's Second Law is to do a calculation. Suppose I throw a 1 kg ball straight up in the air — this makes things simpler as we only have to worry about one quantity, height — and then catch it when it comes back down. We'll click a stopwatch on at the moment the ball leaves my hand, and we happen to know (perhaps by analyzing a film afterwards) that at that moment it was traveling upwards at 20 ms⁻¹ (meters per second — but, as before, the units don't matter. It could just as easily be a 1 lb ball travelling at 20 feet per second).

Once the ball is in the air, there's only one force acting on it: gravity. We know from previous experiments that, close to the surface of the Earth, gravity accelerates everything by about –9.8 ms⁻² (meters per second per second). So we have

$$F = 1 \times (-9.8) = -9.8 \, \mathrm{kg\,m\,s^{-2}}$$

This unit of measurement is also called a "Newton." So far, this isn't very impressive. However, let's rewrite the equation with a dash of calculus:

$$-9.8 = 1 \times \frac{d^2 h}{dt^2}$$

Here, I've invented a function h that tells us the height of the ball at time t. After all, acceleration is the rate of change of speed, which is itself the rate of change of position, so acceleration is the second derivative of position. This, then, just says F = ma in more fancy notation, expressing it as a "differential equation" — an equation with a derivative in it (see The Fundamental Theorem of Calculus, page 26). Thanks, in large part, to F = ma, differential equations are an almost universal part of the language of physics.

We can now transform this equation by integrating both sides. It turns out (you may have to trust me on this) that after one application of this method we get

$$-9.8t + v = \frac{dh}{dt}$$

where v is some unknown value. In fact, though, we know what v is: when the stopwatch starts, t is zero and the speed is 20 ms⁻¹, so we need dh/dt to equal that. That's easily achieved by putting v = 20:

$$-9.8t + 20 = \frac{dh}{dt}$$

We now have a little gadget that can quickly calculate how fast the ball is moving at any given time. We can even say that the ball stopped approximately 2 seconds after it left my hand: common sense tells us that this is the moment when it reached its maximum height, stopped going up and started coming down again. How high did it get? Time for another integration:

$$-4.9t^2 + 20t + s = h$$

where s is another unknown value — except, again, we know what it is. When the clock starts, t = 0, so the height h must be the height of my hand as the ball leaves it. Let's say that's 2 m:

$$-4.9t^2 + 20t + 2 = h$$

So, how high did the ball get? Well, we know it reached its maximum height at about t = 2 seconds,

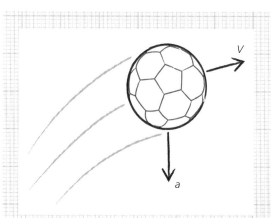

The ball's velocity vector points in the direction it's going; the acceleration from gravity pulls it straight down.

A Mirror up to Nature: Science

so that height was

$$-4.9 \times 2^2 + 20 \times 2 + 2 = 22.4 \text{ m}$$

Many forms of wrestling involve applying force to accelerate your opponent while resisting being accelerated yourself.

A little more algebra will tell us that I'll catch it again about 4 seconds after it left my hand. We started with pretty limited information about the ball's motion; now we know a lot more.

What's more, whether I throw the ball hard or gently, on Earth or on the Moon where gravity is different, integrating $F = ma$ should give me equations that have the same general outline, just with some different numbers plugged into them. In particular, if I plot the height of a thrown ball against time, I should always get an equation that looks like $-At^2 + Bt + C$, which is a parabola (see Conic Sections, page 16). The same observation applies to artillery shells and satellites.

Notice that our mathematical model of the ball isn't perfect. It didn't allow for air resistance, for example, or spin, or the fact that the effect of gravity isn't precisely the same at the bottom of the ball's journey as it is at the top. One great thing about Newton's system is that it turned out to be highly adaptable to things like this: we can tweak our models to give us the level of accuracy we need for whatever we're using them for. At least, that's true until we reach speeds or sizes that require us to consider relativistic or quantum factors (see $E = MC^2$, page 88), but in many practical situations Newton's physics still does a remarkably good job.

Force, mass and acceleration — from any two you can always find the third one. Calculus can translate acceleration into velocity or position, too. This is the cornerstone of Newtonian physics.

Universal Gravitation

The first modern theory of gravity, not superseded until Einstein and still widely used.

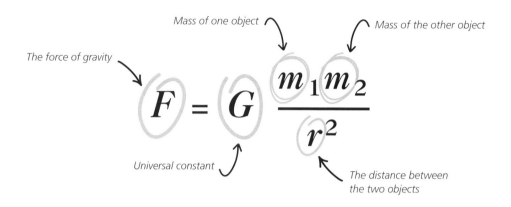

Mass of one object

Mass of the other object

The force of gravity

$$F = G \frac{m_1 m_2}{r^2}$$

Universal constant

The distance between the two objects

Because the force of gravity is less on the moon, golf balls don't get dragged down so fast and can fly further, as astronaut Alan Shepard showed in 1971.

What's It About?

Huge progress was made in cosmology in the 16th and 17th centuries, but by Isaac Newton's time (the late 17th century) the whole subject looked like a bit of a mess. Yes, a lot of new and improved ideas had come on the scene, but many people felt that the harmony and logic of the old system had been lost. "Natural philosophers," as they called themselves, could now make better predictions about the heavens, but what were they based on? The laws of nature, it was felt, really ought to have a sort of underlying simplicity about them.

Newton's *Principia Mathematica* (1687) was important in a number of ways, one of which was this: he effectively proposed that physics should be built around the single key idea of force, and that the workings of the cosmos had, at their heart, a force called "gravity."

There had been theories of gravitation before, but this one sought to make it the foundation of everything else: a single principle to bring order to nature once again. The truth was that nobody, including Newton, had much idea of what this force might be or how it could "act at a distance," as if by magic.

A Mirror up to Nature: Science

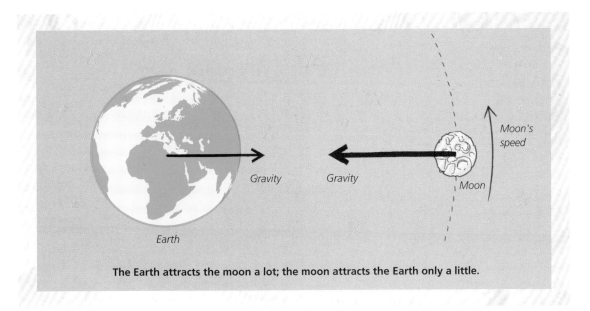

The Earth attracts the moon a lot; the moon attracts the Earth only a little.

In More Detail

Gravity is a force that tends to pull two objects towards one another. In fact, Newton had no idea how gravity might work or even whether it really existed, but adding this mysterious force gave his model great power and simplicity. His famous equation expresses this force in terms of their masses and how far apart they are. Here, gravity has nothing to do with what the objects are made of, where they are, whether they're moving or still, and so on. This makes it clean and simple, and it agrees well with our observations of happenings on the Earth and, especially, in the night sky.

Three things are especially worth noticing about this equation. The first is that the masses of the two objects are multiplied together. This means that a small increase in mass will tend to cause a relatively large increase in the force. The second, related to this, is that the force is divided by the square of the distance between the objects: it's an example of what physicists call an "inverse square law." This means that as the objects get further apart, the effect of gravity drops off *dramatically*.

The third thing to notice is that the equation includes a constant value, G, that doesn't depend on how big the objects are or where they are in relation to each other. In fact, G doesn't depend on anything at all: it is, as far as we know, a "universal constant" that has the same value in every physical setup everywhere in the universe.

Today, we consider gravity one of the four fundamental forces of the universe. All other forces can, in theory, be described in terms of them. One of the others is electromagnetism, which wasn't well understood until the 19th century (see Maxwell's Equations, page 92); the remaining two act only at tiny distances within atoms, so they're far removed from our ordinary experience.

Gravity is one of the most fundamental forces in physics; this equation shows how it depends on the masses and distances of the bodies attracting each other.

Conservation of Angular Momentum

A basic law about spinning things that govern the behavior of (among other things) figure skaters, tightrope walkers, flywheels and neutron stars.

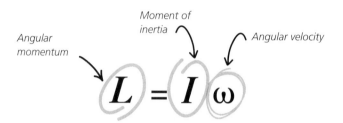

Angular momentum → *Moment of inertia* ↘ ↙ *Angular velocity*

$$L = I\omega$$

Christiaan Huygens, a Dutch scientist of the 17th century, demonstrates his most famous invention, a pendulum clock.

What's It About?

If you have a swivel chair, the best way to understand this equation is by performing a simple physical experiment. Sit on the chair and spin around: if the bearings are well oiled it will seem to spin for a few seconds before friction slows it down. Now do it again, but after a second or two stick your arms out, with something fairly heavy — a book, perhaps — in each hand. Notice that you suddenly slowed down. Now do it again, but this time start with your hands out and bring them in to your chest after a second or so; this time your rotation speeds up. Experiment with sticking your hands out and pulling them in and you should find you can control the speed of the spin quite effectively.

What's happening here is that the angular momentum of the books in your hands is being conserved: when they stick out, they're traveling further in a single rotation, so the speed drops to compensate. When they're held to your chest they

A Mirror up to Nature: Science

don't travel so far, so they can go faster. When you try it, it may seem a tiny bit miraculous that the chair "knows" to speed up or slow down to compensate for how you're holding your hands, which aren't touching the chair at all!

Why Does It Matter?

Angular momentum is a fundamental property of things that rotate. You might say that, second only to going in a straight line, some sort of rotation is just about the most popular form of motion in the universe.

The fundamental idea of angular momentum emerged in the late 1600s from the study of clocks and pendulums by the Dutch scientist Christiaan Huygens and others. Pendulum-like motion is used in many pieces of mechanical technology besides clocks, and an understanding of angular momentum was important to all the flywheels and governors that made the industrial revolution possible. The back-and-forth swing of a pendulum might not seem to have much in common with steady circular spinning — and certainly momentum isn't conserved in an ordinary pendulum, since it stops at the top of each swing — but the two are very intimately related (see The Damped Harmonic Oscillator, page 78).

Human beings who make a habit of rotating can feel and use the effects of the law very immediately. Gymnasts and acrobats who perform movements such as backflips always tuck their limbs in tight so that the rotation in the air happens quickly, ensuring they get a full rotation in before they land; after all, it's a terrible idea to do half a backflip. Figure skaters use the swivel-chair principle to control their speed in a spin, and high-board divers, hammer or discus throwers and bat, club and racquet swingers of all kinds make use of it, too. More exotically, it's the reason why a tightrope walker holds a long pole and a bull-rider at the rodeo puts out a hand to keep balance.

It isn't just humans or machine parts that use the Conservation of Angular Momentum on a regular basis. Animals with tails often use their angular momentum to perform feats of balance or agility. The famous ability of cats to land on their feet by twisting in the air as they fall depends on it, but please don't put this to the test on your local feline; though impressive, this maneuver doesn't have quite as good a success rate as legends would have you believe.

Many objects out there in the cosmos rotate, too. The most dramatic of these are the stars called

An ice skater pulls in her arms, decreasing her moment of inertia. Her velocity automatically increases to conserve angular momentum.

Conservation of Angular Momentum

"pulsars": unimaginably dense objects that spin at fantastic speeds. This happens because a large star like our sun, rotating relatively slowly, will eventually collapse down to a tiny fraction of its size to form a neutron star. That's like pulling in your hands on a spinning office chair, but on a truly astronomical scale.

At the other extreme, all elementary particles have a property called "spin," which is a kind of angular momentum. In the quantum world, though, we might look at this as more of an analogy than a literal statement, because particle-wave duality means it's not quite right to think of an electron, say, as a little ball spinning in space (see The Schrödinger Wave Equation, page 104).

In More Detail

Momentum represents how hard it is to stop a thing from moving (see Newton's Second Law, page 56). Linear momentum — that is, the momentum of something moving in a line, rather than spinning in a circle — is just mass times velocity. This makes sense: the factors that make stopping an object in its tracks difficult are how heavy it is and how fast it's going. Put another

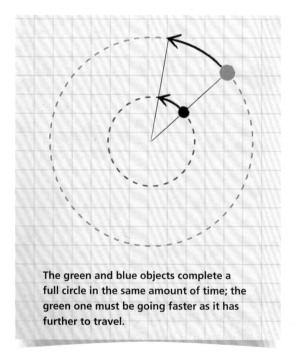

The green and blue objects complete a full circle in the same amount of time; the green one must be going faster as it has further to travel.

way, it hurts more to get hit by a car going at 30 km/h (20 mph) than by a paper airplane going at the same speed, and it's worse still if the speed increases. Notice, too, that it's a multiplier: getting hit by a car going twice as fast is much worse than getting hit by a paper airplane going twice as fast because the mass being multiplied is so much bigger.

Angular momentum uses the same basic ideas but it's a shade more subtle. First off, we need to measure our velocity with angles, that is, in something like degrees per second rather than meters per second. Someone spinning at 60° per second would complete a full turn (360°) in a sedate 6 seconds. This angular velocity is represented by ω in the equation.

The mass part is more tricky, too. Consider the situation with the swivel chair again. Suppose someone sits on the chair and spins with their arms out. Now imagine they do the same, at the same angular speed, but with long poles sticking out. Which would hurt more, getting whacked by the hands or the ends of the poles? It's intuitively obvious that the ends of the poles are moving faster through space. A spin cycle takes a fixed time but the ends of the poles have further to go to complete it, meaning they pack more of a punch and take more force to stop. This aspect of the "difficulty of stopping" is measured by the moment of inertia (I) and that, times the angular velocity, which represents another aspect of the same difficulty, gives us our angular momentum. The motion of the planets around the sun is governed by this principle, too. They move not in perfect circles, but ellipses (see Kepler's First Law, page 52). This means that during some parts of the orbit, their angular velocity is larger than usual. It's almost as if the Earth is on the end of a piece of string being spun around by the sun — when the string gets pulled short, the Earth goes faster. This principle applies to all orbiting bodies, everywhere in the universe, as long as external forces aren't having any significant effect on them.

Conservation laws generally only apply to "isolated" or "closed" systems — that is, physical setups that aren't subject to interference from the

A Mirror up to Nature: Science

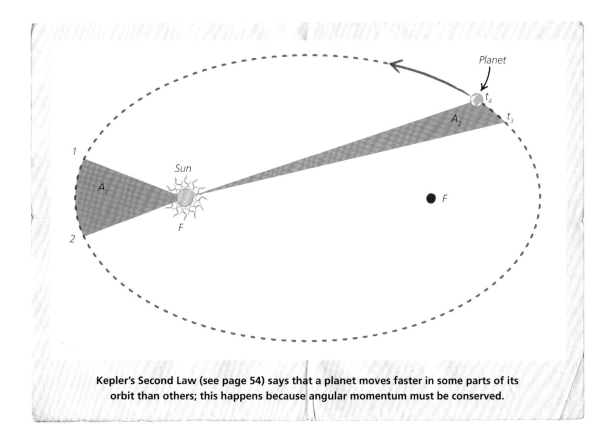

Kepler's Second Law (see page 54) says that a planet moves faster in some parts of its orbit than others; this happens because angular momentum must be conserved.

outside. Notice, for example, that the swivel-chair experiment doesn't work if someone is randomly giving the chair a push or slowing it down while it's spinning — in that case, all bets are off as to what might happen.

You might have heard of other conservation laws in physics that apply to mass, energy, linear momentum, and so on. In 1915, the German mathematician, Emmy Noether, proved that laws of this type are always the result of a certain kind of symmetry within the equations governing a physical system. This is a very abstract, general result. It has led to the discovery of more conservation laws since then, including the gauge symmetries of particle physics, and in a sense it has given us a deeper idea of what a conservation law in physics really is. Incidentally, the absence or "breaking" of an expected symmetry of this kind was intimately tied to the conjectured existence of the Higgs Boson elementary particle, which was confirmed by experiments at CERN in Switzerland, almost half a century later.

Conservation laws abound in physics, and usually tell us something deep about what's happening, and circular motion is prevalent enough to make this one especially important.

Conservation of Angular Momentum

The Ideal Gas Law

This simple equation brings together temperature, pressure, volume and mass to explain many real-life phenomena, including pressure cookers and hot-air balloons.

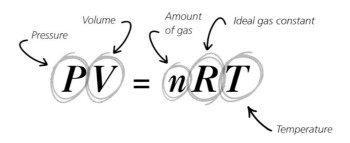

$$PV = nRT$$

Pressure — Volume — Amount of gas — Ideal gas constant — Temperature

What's It About?

Maybe the Ideal Gas Law isn't as amazing or impressive as many of its peers in this book. Perhaps, in comparison to them, it even looks a bit drab and shabby. It's here for two reasons, though. The first is that it packs a lot of information into a neat little package; the second is that gases are pretty much everywhere.

Perhaps you already know, at least in common-sense terms, that there's some relationship between the pressure, temperature and volume of a body of gas. If you fill a balloon with hot air it will deflate a bit as the air cools down, because cool air takes up less volume. If you squeeze a balloon (decreasing its volume) you might burst it, since you've also increased the pressure of the air inside. We're used to similar sorts of interactions when cooking. This equation captures the principle that makes pressure cookers work — the same reason a pan of water boils more quickly with a lid on top, since that too causes pressure to build up, which in turn raises the temperature. These are the principles behind, among other things, the steam engine — and we all know how significant an invention that was.

In More Detail

It's hard to see what all these terms mean as long as you think of a gas as a fuzzy cloud of stuff. "Temperature" and "pressure" in particular look like mysterious qualities the stuff has, like the "intelligence" or "willpower" a person might have. Are they clearly defined at all?

Two containers of gas, one smaller than the other. The gas particles are represented by little balls bouncing around. In the smaller one the balls are more tightly packed.

A Mirror up to Nature: Science

François Pilatre de Rozier made an early ascent in a captive hot air balloon in Paris, on October 11, 1783.

James Eckford Lauder's painting of James Watt working on the steam engine.

It makes things much easier if you imagine a gas as a bunch of tiny, weightless ping-pong balls (atoms) bouncing around. Think of a balloon: the temperature of the air inside depends on how fast, on average, the atoms of air are whizzing about. The pressure depends on how often they bounce against the stretched skin of the balloon.

Now the Ideal Gas Law is almost obvious. Of course, if you heat up the air, meaning that the atoms are whizzing about more quickly, you'll get more collisions with the skin (and so more pressure). Of course, if you make the balloon smaller, but keep the same amount of air in it, the atoms will bump into the skin more often (again, more pressure). If you want to keep the atoms whizzing around at the same speed (the same temperature) but reduce the number of collisions with the edges (the pressure), then it makes sense to move the edges away (increase the volume).

We can rewrite the Ideal Gas Law as

$$\frac{P\,V}{n\,T} = R$$

where R is the Gas Constant, which is a universal physical constant. That is to say, as far as we know, R has the same value everywhere in the universe. It's closely related to Boltzmann's Constant (see Entropy, page 74).

You might be surprised at first that the Ideal Gas Law, which seems to be about rather earthly things, is true everywhere in the universe. When you think of it as a simple statement about the statistical movement of tiny particles, though, it's easier to see why that might be so.

The basic fact about how a gas works: compress it and the temperature of the gas rises, heat it and it expands, remove some and the pressure goes down.

The Ideal Gas Law

Snell's Law

The law that tells us how to control the direction of a beam of light.

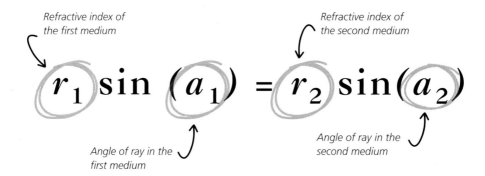

Refractive index of the first medium

Refractive index of the second medium

$$r_1 \sin(a_1) = r_2 \sin(a_2)$$

Angle of ray in the first medium

Angle of ray in the second medium

What's it About?

You've probably seen the following trick: take a glass of water and put a pencil halfway into the water. Depending on how you look at it, the pencil will appear to be either bent or broken at the water's surface. This happens because of refraction: the fact that when light moves from air to water it changes direction a bit.

You may also notice something else about this situation, although it's so common it may pass you by. When you look at the surface of the water, you'll probably see light glimmering on it. The water isn't producing light: it's reflecting some of the light that comes down onto it from above.

Here we have two different but evidently related phenomena — refraction, in which light rays get bent or diverted a bit, and reflection, in which they are completely bounced off in another direction. Snell's Law describes both of these situations.

Refraction of light makes many optical technologies work, from the lenses in your glasses to the laser in your CD or DVD player. As for reflection, its principles are used in car headlights, the Hubble Space Telescope and the fiber optics that form the backbone of the internet.

What's more, many other things besides light

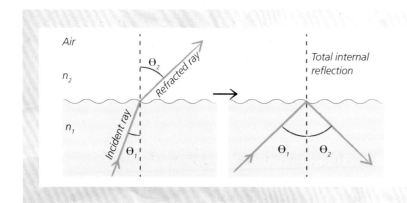

Air

n_2

n_1

Θ_2

Refracted ray

Incident ray

Θ_1

Total internal reflection

Θ_1

Θ_2

Whether a ray of light escapes from one medium to another or bounces off the boundary, Snell's Law tells us exactly how it will happen.

A Mirror up to Nature: Science

Snell's Law describes how the refraction of light makes a pencil in a glass of water appear to bend.

obey Snell's Law, at least closely enough for it to be useful, including sound and radio waves. Indeed, if you're any good at pool, snooker or billiards you must have a pretty good practical knowledge of it. In common with several other equations in this book, this information also enables us to build realistic simulations of physical phenomena for video games and movies as well as more serious tools for research and training.

In More Detail

Snell's Law is a conservation law (see Conservation of Angular Momentum, page 62); that is, it says

that although something changes, something else stays the same. What changes here is the medium the light is passing through. Air and water are different media, for example, and each has different physical properties that affect the way light moves through it.

The quantities involved are rather simple: the angle the light's traveling in before and after it passes into the new medium and the refractive index of each medium. The angle is measured relative to a line standing straight up, at right angles to the surface between the two media at the point where the light hits it. We use the sine function to factor out any cases where the angle measurement happens to include one or more extra "full turns" of 360°, which don't make any real difference (see Trigonometry, page 14).

The refractive index is a number we attach to the material the medium is made of. This is another ratio, like the sine of an angle, meaning it represents a proportion rather than a quantity, and will be the same regardless of what units of measurement you use (see The Cross-Ratio, page 118). It's calculated as the speed of light in a vacuum (which is a constant) divided by the speed of light in the medium in question: in other words, it's the amount that light is slowed down when it tries to pass through the medium. Think of how you might be slowed down when crossing from a nice tarmac road to a muddy field and you'll get an insight into how light feels when it goes from air to water. Snell's Law says that although the refractive index can change, and the angle of the light beam can change, the proportional relationship between them stays the same.

Reflections and refractions are at the heart of the science of optics, and they're governed by an elegant relationship of trigonometric ratios.

Brownian Motion

The random-looking movements of tiny particles provide a surprising model for heat flow and financial markets.

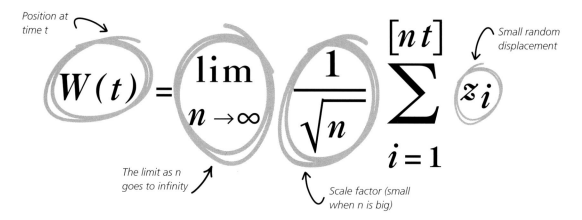

Position at time t

$$W(t) = \lim_{n \to \infty} \frac{1}{\sqrt{n}} \sum_{i=1}^{[nt]} z_i$$

Small random displacement

The limit as n goes to infinity

Scale factor (small when n is big)

What's It About?

In 1827, the biologist Robert Brown was looking at pollen grains under a microscope when he noticed something odd about their movement. They were floating in water but, instead of gently drifting around like a stick floating on a pond, they tended to move in a jerking, random-looking fashion. This observation was repeated by others, but couldn't be explained. It wasn't until 1905 that Einstein came up with an explanation, but a lot had happened in physics since then that Brown himself couldn't have anticipated.

To get an idea of Brownian Motion, imagine a tightly packed crowd at a concert or sporting event all waving their hands in the air. You throw a beach ball into the crowd and the waving hands will bat it around; each time the ball touches a hand, it gets flicked off in a new, effectively random direction. The ball is undergoing a "random walk" (see right). If you imagine a vast crowd of people and watch the ball's motion for a long time, the effect is a lot like Brownian Motion.

This kind of motion is a strange thing: a continuously wriggling, twitching jiggle that looks like nothing we see in the everyday world. Natural processes tend to be smooth, like the curving path

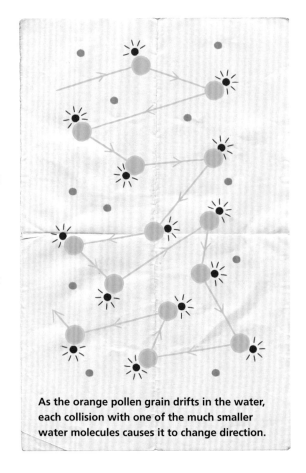

As the orange pollen grain drifts in the water, each collision with one of the much smaller water molecules causes it to change direction.

A Mirror up to Nature: Science

of a falling leaf blowing in the wind, but those particles of pollen seem to have an angular motion that jumps from one direction to another without ever settling down.

Not surprisingly, then, it needs some rather strange mathematics to make a good model of it. Ordinary calculus, the bread-and-butter of physical science, has difficulty getting hold of Brownian Motion because of its spiky changes in direction; in fact, a whole new system of calculus had to be invented to cope with it.

Why Does It Matter?

The mathematical model of Brownian Motion took a while to develop and when it did it represented a rather new kind of object. As a result, variations of it have been used to attack problems that other techniques couldn't quite get a grip on. Biologists have used it to make improved models of the behavior of birds, fish and insects moving in large groups; it's been used to enhance noisy digital signals, including medical ultrasound images; it's also frequently used to model financial asset prices and inform trading decisions.

What's more, the mathematics of Brownian Motion is weird enough to be interesting in its

Robert Brown was perplexed by the jerky motion of pollen grains he watched under a microscope.

own right. In the 19th century the calculus devised by Newton and Leibniz, which was used everywhere in the sciences, had come under attack from a series of strange, artificial examples that seemed to defy some of its most basic assumptions. Brownian Motion is an example of such a function, and the fact that it's a very good model of many natural phenomena reminds us that the gulf between "pure" and "applied" mathematics isn't as big as we might think.

In More Detail

John, I'm afraid, has had rather too much to drink and is attempting to walk home. To do this he needs to walk in a straight path across a large field. He does his best, but each step he takes, as well as going forward, involves a random stagger to the left or right so that, from above, his path looks like a jagged line (see illustration on page 73).

If you want to make this example more precise, imagine tossing a fair coin (see The Uniform Distribution, page 162) and shoving John gently to the right if it comes up tails, to the left if it comes up heads. This is what's called a "random walk." As time goes on, John moves steadily forward (from one side of the field to the other), but he also weaves sideways across the field.

When John gets to the other side of the field he might get lucky and reach the gate, but he could just as well end up stuck in the hedge. So one question we might ask is: what are the chances he'll make it to the gate? Though it's not completely straightforward, we get an answer from basic probability theory.

In fact, we should (in a slightly technical sense) expect John to make it to the gate, because his leftwards staggers and rightwards staggers ought to balance out, on average. Naturally, a particular walk could land him in any number of hedge-related predicaments, but if we had to make a bet on his final position we should bet on the gate as that's where he'll wash up most often. We can also ask, for example, how likely it is that John will end up within 3 meters (or 3.3 yards) of the gate, more than 10 meters (or 10.9 yards) away and so on — again, we can get sensible answers from our model.

John's path has what probabilists call the "Markov Property," which says that at every moment during the walk across the field, the next step depends only on where he is, not on what's happened previously (see Fibonacci Numbers, page 22). Essentially, the left-right staggers have no memory of what's gone before. In real life John might start veering to the left and develop a bit of momentum, making more steps to the left more likely; this wouldn't be a Markov process any more. This property is important because it greatly simplifies things.

So far, so good, but we don't yet have Brownian Motion. Suppose John takes n steps to cross the field — that is, n steps forward, each accompanied by a stagger to the left or right, so the total effect is a diagonal movement. To turn this into Brownian Motion, we start increasing n — that is, we make John take smaller steps. Suppose he takes steps half the size he did before; then n doubles and his path acquires twice as many left-right staggers. Now suppose we repeat this process over and over, doubling n each time and watching what happens as n gets larger and larger.

What's surprising is that the answers to questions about this process — such as the

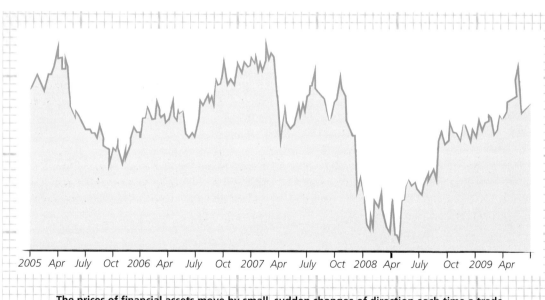

2005 Apr July Oct 2006 Apr July Oct 2007 Apr July Oct 2008 Apr July Oct 2009 Apr

The prices of financial assets move by small, sudden changes of direction each time a trade is made. As a result, their long-term behavior looks a lot like Brownian motion.

A Mirror up to Nature: Science

chances that John will end up in a certain part of the far boundary — settle down to stable values as n gets larger. This encourages us to look at the limit of John's walk as n goes off to infinity (see Zeno's Dichotomy, page 18), which results in the mathematical object called the Wiener Process.

We're not helped much any more by imagining a drunk staggering across a field: that's too mundane a picture for such an exotic object. Let's go back to the pollen grain, which is tiny and very light. When it floats on the water, it's actually being buffeted by millions of water molecules, which are whizzing around and bouncing off each other in such a complex way that it's effectively random (see The Ideal Gas Law, page 66). Each time a molecule hits the pollen grain, which happens thousands of times a second, it gives it a minuscule nudge. The overall effect of these nudges is the wandering motion that Robert Brown observed through his microscope.

Now, it has to be said that there's a big difference in principle between millions of tiny nudges and what the Wiener Process gives us, which is an infinite number of infinitely small nudges — an idea that, on paper, doesn't even seem to make any physical sense. Though these infinities can be wrestled into mathematical order with a bit of extra work, they only approximate the physical situation. Still, they do it very well, and the Wiener Process itself turns out to be important in other, sometimes surprising areas (see The Heat Equation, page 80). One of those is high finance. In 1900, Louis Bachelier, a student of the great Henri Poincaré, published his Ph.D. thesis as *The Theory of Speculation*. In it, he analyses the movements of prices on the Paris stock exchange using the then-new theory of Brownian motion. Bachelier's ideas didn't gain much recognition until the 1960s, but as the computer age dawned Brownian Motion became a popular way to

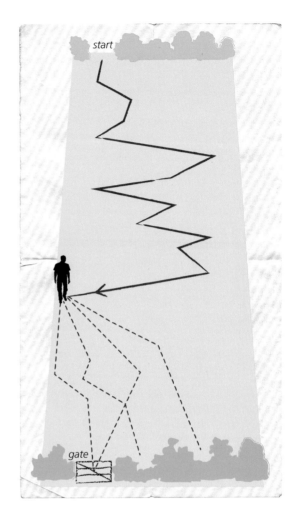

What are the chances John will make it to the gate? The answer depends only on where he is now, not on the path he took to get there.

simulate share price movements in the future, giving predictive power to otherwise merely descriptive models like the Black-Scholes Equation (see page142).

Random walks model unpredictable behavior that evolves over time, with each change being independent of the previous ones. Brownian motion takes this to its ultimate conclusion.

Brownian Motion

Entropy

The Second Law of Thermodynamics — it's the reason your coffee eventually gets cold and may predict the final fate of our universe.

Entropy

Boltzmann's constant

$$S = K \ln(W)$$

Number of states the system can be in

What's It About?

The slogan for the Second Law of Thermodynamics is "Entropy never decreases." What does this mean? Often entropy is described as the amount of disorder in a system. Consider a junkyard in which the parts of various vehicles are piled up more or less neatly: if a violent storm were to blow through it, we might expect to find these nice piles of parts scrambled up and strewn about the place. We'd be very surprised if the storm had assembled them into working cars and trucks once again. That's (in part) because we expect natural processes to reduce the amount of organization or pattern in a system and to spread things around evenly. This intuition even

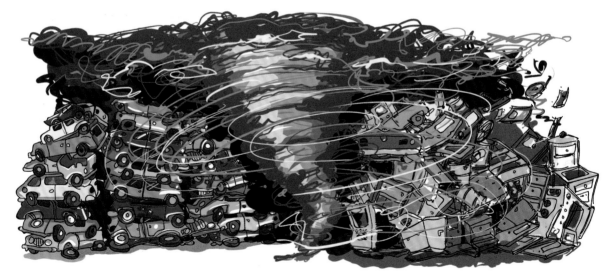

Is the tornado moving right-to-left or left-to-right?
One of those is more likely than the other.

A Mirror up to Nature: Science

has a theological application: it lies at the heart of many of the famous "arguments by design" for the existence of a divine creator.

Here's a less fancy example, right now, the air in the room you're in is made up of billions of tiny molecules flying around, bouncing off each other. The movement of any one of them is so complex that it's effectively unpredictable. It's conceivable that this movement could result in all the molecules being in one corner of the room, leaving you gasping for air — possible but hugely unlikely. You'd have to put some energy into the system to make it behave like that, and probably to keep it like that too; left to its own devices, a room with all the air in one corner will quickly go back to the normal state of things. That's entropy increasing, isn't it?

The trouble with this picture is that it isn't very precise. What do we mean by "disorder," "randomness" and "structure"? Are these things we can measure, and so speak accurately about them increasing or decreasing? On the face of it, the Second Law seems more like a qualitative statement than a quantitative one; that is, one that deals in potentially vague general properties of a system rather than hard numbers. The tricky part of all this, then, is to turn our intuitions about entropy into something more concrete.

Why Does It Matter?

It's not much of an exaggeration to say that entropy and its non-increasingness affect every part of your life, including the fact that it's finite, for eventually entropy wins out over the wonderful organization of our biology, too. Perhaps it's a small comfort to know that the same principle presides over the deaths of stars as well.

Returning to more everyday matters, consider placing something hot next to something cold. This is a low-entropy system — intuitively, it's highly organized and very unlikely to happen spontaneously. Similarly, all the heat in the room you're in right now isn't just going to suddenly flow into one spot, causing your coffee to boil while freezing everything else. Instead, what actually happens, every single time, is that the heat from

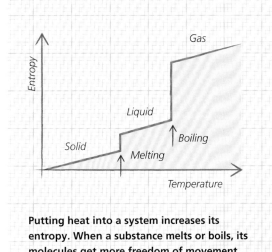

Putting heat into a system increases its entropy. When a substance melts or boils, its molecules get more freedom of movement and its entropy leaps up.

your coffee flows out and dissipates through the room.

As entropy increases, heat passes from the hotter object to the colder one and warms it up. So, it is that the Second Law of Thermodynamics makes cooking possible, along with engines and industrial processes that use the transfer of heat (which is nearly all of them). It comes at a cost, though: the Second Law also puts a limit on how efficient such a machine can be and rules out the existence of a perfect engine that converts 100% of the energy it consumes into work.

All these describe one-way, irreversible processes. In classical mechanics, we're used to being able to run time backwards and get a perfectly reasonable-looking interaction. For example, in theory if you watch a game of pool In reverse all the physics is absolutely fine — you just have to imagine that the energy that moves the balls is coming from some mysterious force acting on the table rather than from what the player ate for lunch. (In practice, this isn't actually true: real-world events are too complicated to be perfectly reversible.)

Entropy isn't like this: it puts an "arrow of time" at the heart of physics, a sense that the universe is something other than a piece of clockwork.

With entropy, we see the universe as heading in a distinct direction. The end-point might be the heat death of the universe, in which all matter and energy is equally spread out in an undifferentiated haze across the entirety of space, so maybe it's not a cause for much optimism on the grandest scale.

In More Detail

The first thing to say here, is that the real thermodynamic slogan is: "Entropy never decreases in a closed system." A closed system is one that has nothing happening to it from the outside. The

An externality usually explains violations of the Second Law. Here, Hieronymus Bosch depicts God separating the heavens from the Earth at the moment of creation.

Earth, for example, certainly isn't a closed system because the sun is pouring down energy on it from the outside. Similarly, the junkyard isn't a closed system if a team of engineers roll up and start building stuff out of the parts: that's an external effect that would fully explain the parts being

A Mirror up to Nature: Science

rearranged into working machines again.

Certainly, then, entropy often decreases when energy is being put into a system. The claim of the Second Law only holds without qualification if we look at a big enough, perhaps cosmic scale. On that scale, we're referring to average entropy rather than specific, much smaller-scale events. What's more, entropy could decrease by chance on rare occasions; the point is that the probability of such a thing happening is vanishingly small. If you'd like an analogy: when you drop a pencil it theoretically could land on its point and balance there. It just never, ever does.

There are more available states outside the balloon than inside. The air tends to rush out of the hole in accordance with the law of increasing entropy.

Within a closed system the official definition of entropy uses the logarithm of W, the number of states the system can be in. This "number of states" is where the action is, and understanding it gets you most of the way to understanding entropy; the logarithm and Boltzmann's Constant k are really just scaling devices to make the definition work better (see Logarithms, page 36). So, what is this number W?

Think of a balloon full of air. From the standpoint of classical physics, the state of the inside of the balloon as a system is given by the position and velocity of each air molecule, all taken together. That's all well and good, but there are an awful lot of them and they're all moving about in a very complex way. Although, in theory, we could set up equations to describe them, in practice that would be much too difficult. Instead we have a general "statistical" picture of how they're behaving — very likely they're pretty evenly distributed throughout the inside of the balloon and, although they'll be moving at a variety of speeds, there's nothing to suggest that the average speed will start going up or down.

Now, suppose we put a hole in the balloon with a pin. It could be that the air molecules carry on doing what they were doing and ignore the hole, but that's highly unlikely. More likely the air will rush out of the hole and into the surrounding area. What makes this more likely? The Second Law of Thermodynamics. For if the air molecules rush out, they'll be in a bigger space that offers lots more states for them to be in — the total entropy of the system has increased.

This is the bit of classical physics that's not like the others: entropy embeds an "arrow of time" in the universe and encourages us to think about problems statistically.

The Damped Harmonic Oscillator

From springs to synthesizers, this versatile model has found its way into a huge amount of technology.

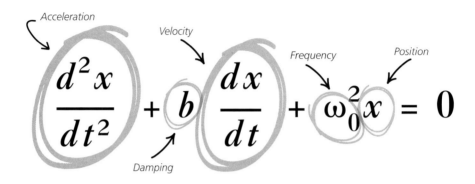

Acceleration

Velocity

Frequency

Position

$$\frac{d^2x}{dt^2} + b\,\frac{dx}{dt} + \omega_0^2 x = 0$$

Damping

What's It About?

Put a plastic ruler over the edge of a table and "twang" the end. It will vibrate up and down in a rather satisfying manner, the vibrations getting smaller and smaller until eventually they peter out. This is a Damped Harmonic Oscillator. It's an "oscillator" because it changes smoothly from one state (all the way up) to another (all the way down) and back, over and over again. It's "harmonic" because the way it changes follows the pattern called a sine wave (see Trigonometry, page 14). It's "damped" because it loses energy as it vibrates, causing it to slow down and stop.

Damped Harmonic Oscillators are pretty common objects. Pendulums, springs, balls, piano strings, children on playground swings and sound waves are just a few examples. The suspension system of a car is deliberately heavily damped to stop you from bouncing around too much after going over a bump. In early musical synthesizers (and still sometimes today) the electronic waves that produce the sound are harmonic oscillators

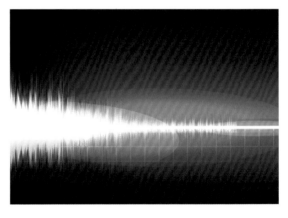

Sound is a wave, but many sounds fade away rapidly due to damping.

that can have damping applied to them to create the effect of (for example) a plucked violin string rather than a note sustained indefinitely by the bow. We often see some sort of damping or decaying effect because of things like friction that slowly take energy out of the system as it moves.

A Mirror up to Nature: Science

In More Detail

The equation is derived quite directly from $F = ma$ (see Newton's Second Law, page 56). It involves all three quantities of acceleration, velocity and position, along with numbers representing the amount of damping and the frequency of the oscillation.

At the top of a vibration the velocity momentarily becomes 0, since it's changing direction. At this moment, the equation becomes

$$x'' + w^2 = 0$$

so the acceleration has a negative value, meaning it's pushing down on the ruler. Similarly, at the bottom of the motion the acceleration is pushing up instead, and these are the effects that cause it to change direction. So, in a sense, the tension in the ruler is always pushing it back towards the middle.

When it's right in the middle, on the other hand, there's no tension, but the damping force is still in effect. If we take the damping out (that is, put $b = 0$), there's actually no acceleration at all at the instant when the ruler is perfectly straight; it's just its momentum that carries it past that point to create another oscillation. When there's no damping we say it's engaged in "simple harmonic motion," but in real life there's almost always a damping effect of some kind.

In fact, everyday things don't always oscillate in the nice, neat way the Damped Harmonic Oscillator does either. For example, a violin string does something much more complex, as does a flute, and that's a big part of what makes the two instruments sound different. To understand this we need more sophisticated methods (see The Wave Equation, page 84).

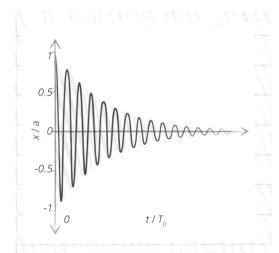

The damped harmonic oscillator at work: it vibrates up and down like a sine wave but the vibrations die away over time.

We can get a bit more complicated by adding "driving" as well as damping into our model. This is effectively an extra part of the system that gives it a "kick" of some kind. You can simulate this with your ruler by flicking it once every 2 seconds (say) and seeing what happens. The behavior is a bit more complicated: after each flick, the damping starts to slow it down, but then the flick drives it back up again. Notice that if you time the flicks right, you can cause the ruler to vibrate much more violently than it was at first.

Dribbling a basketball is like this, as is the slight bounce a bridge experiences when people walk on it. As you may know, soldiers break their synchronized marching step when crossing a bridge so they don't drive the bridge to oscillate dangerously, which has been known to cause structural damage.

Trigonometry, which helps us understand circular motion, also gives us a grip on its close cousin, oscillation. Adding factors for damping and driving into the mix enables it to model many real-life situations.

The Damped Harmonic Oscillator

The Heat Equation

The way heat moves through an object has a close and surprising relationship to statistics and financial simulations.

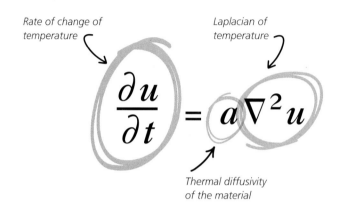

Rate of change of temperature

Laplacian of temperature

$$\frac{\partial u}{\partial t} = a\nabla^2 u$$

Thermal diffusivity of the material

What's It About?

Suppose you point a blow torch at a big, complicated object — a car, for instance. Aim the flame at the driver's door and hold it there for a long time. What will happen? Common sense tells us the spot where the flame's touching the door will get very hot indeed. What else will happen? Experience suggests that the area of the door close to that point will get hot too, even though the flame isn't touching it. On the other hand, we'd be surprised to put a hand on the hood and find the blow torch had made it hot all the way over there.

Intuitively, we understand that heat won't just build up at a point but will tend to flow towards colder parts it's in contact with (see Entropy, page 74). Yet, we also know it doesn't just get immediately dissipated throughout the universe so that no heat can build up anywhere. There must be some kind of pattern to how heat moves in this sort of situation: that's what the Heat Equation captures.

Why Does It Matter?

There are many situations in science and engineering in which heat plays a critically important part. Being able to determine how the intense heat in a nuclear reactor will dissipate can mean the difference between a safe operation and a disaster. It also helps geologists to understand how the Earth's continents formed and to predict the effects of volcanic eruptions, climate change and earthquakes. The Heat Equation describes the behavior of more everyday phenomena, too.

The Heat Equation's close relatives, the Reaction-Diffusion Equations, may explain how some complex patterns, such as zebra markings, arise in nature.

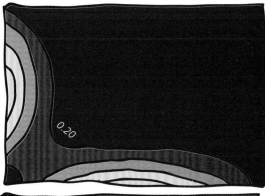

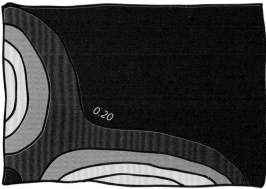

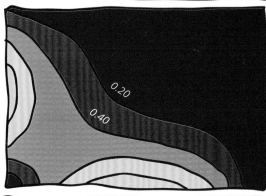

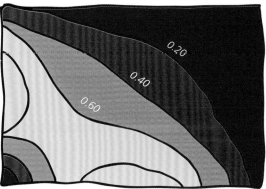

It needn't be about physical heat, either. The function u (see page 82) can be something other than temperature, and as long as the relationship of the Heat Equation holds we can use it as a model and answer questions with it, as if we're using the diffusion of heat through a material as a detailed analogy of the phenomenon we're trying to understand.

This equation makes frequent appearances in biology, where it's often called the "diffusion equation." In the slightly more complicated setting of reaction-diffusion systems, it helps to answer questions about the movement and spread of populations, the process of healing, the growth of cancer cells and perhaps even the way animals like tigers and zebras get their complicated-looking markings. It can also be used to clean up digital images, helping with everything from forensics to astronomical observation.

There's an even more surprising application of the Heat Equation in mathematical finance. Asset prices in a market can be modeled by a random process (see Brownian Motion, page 70). Derivatives contracts that rely on these prices change their own value in complicated ways that turn out to be intimately connected to the Heat Equation (see The Black-Scholes Equation, page 142).

In More Detail

You can think of heat flow as a statistical process involving billions of atoms or molecules in the sheet of metal (or whatever it is) all jostling against each other. The hot atoms vibrate faster, but that means they jostle the atoms next to them more, which means these get hotter too, while the original jostling atoms lose a bit of their energy in the process. In this way we can see the energy being dispersed just as the atoms of air in a balloon of a gas disperse when you burst it. The Second Law of Thermodynamics reminds us that this is a

The Heat Equation describes how heat spreads through a cold slab of metal as the two flames are applied to its edges.

In digital image processing the Heat Equation can be used for everything from forensic or scientific enhancement to artistic effects.

one-way process: the heat that's dispersed in a cooling slab of metal never flows back into a concentrated spot even if that's where it started (see Entropy, page 74).

The Heat Equation contains some symbols that might need a bit of unpacking before it makes much sense. First off, this is about changes in temperature at every point in a three-dimensional space over a period of time. In the equation, u effectively attaches a number, the temperature, to every point in four-dimensional spacetime.

On the left is a derivative (see The Fundamental Theorem of Calculus, page 26) that will tell you, given a point in space and moment in time, how the temperature's changing with respect to time. For example, it'll give you a big positive number if the temperature's going up really quickly, or a small negative one if it's slowly cooling down.

This is wonderfully useful information if you know how to calculate it. You could think of the left-hand side of the equation as a question: "How does the temperature at a chosen point change over time?"

On the right, the a represents the thermal diffusivity of the material: it's just a number that tells us about the way heat flows in this particular sort of material. Being multiplied by it is a monster called the "Laplacian" of u, and this is the secret heart of the Heat Equation. In a three-dimensional

space the Laplacian is defined as follows:

$$\nabla^2 u = \frac{\partial^2 u}{\partial x^2} + \frac{\partial^2 u}{\partial y^2} + \frac{\partial^2 u}{\partial z^2}$$

On the right we're adding up three things, each of which stands for the acceleration of the heat flow in the x-, y- or z-direction at the given point at that moment in time.

It might help to think of a simplified situation: a flat, two-dimensional space such as a thin slab of metal that's being heated in some way. At each point on the slab, at every moment in time, $u(x, y, t)$ gives you the temperature — notice we only have x and y now, not z, because we've thrown away one of the dimensions. Imagine putting this slab on the floor and treating the number (representing temperature) as a measure of height. This creates a sort of ghostly landscape where the tops of hills are directly above very hot places on the slab and the bottoms of valleys are above the coldest ones. Over time, this landscape will gradually change shape as heat dissipates through the slab.

Now, imagine standing at a point (x, y) on the landscape while this is going on. The temperature there is $u(x, y, t)$, where t is the current time. If you look in the x-direction, the slope you see is

$$\frac{\partial u}{\partial x}$$

that is, the rate of change of temperature in the

A Mirror up to Nature: Science

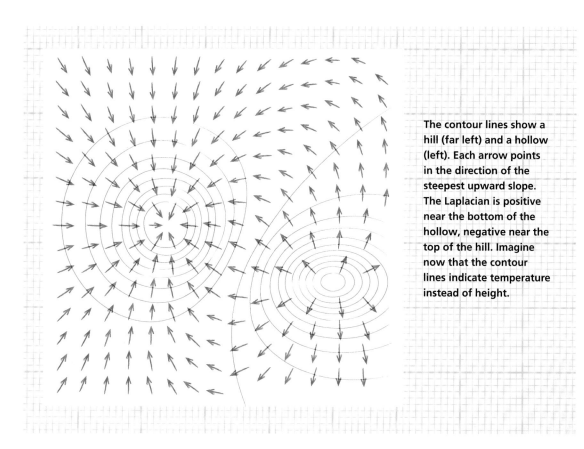

The contour lines show a hill (far left) and a hollow (left). Each arrow points in the direction of the steepest upward slope. The Laplacian is positive near the bottom of the hollow, negative near the top of the hill. Imagine now that the contour lines indicate temperature instead of height.

x-direction. Now, keep watching: perhaps the slope flattens out, or gets steeper? That effect is described by

$$\frac{\partial^2 u}{\partial x^2}$$

the rate of change of the heat flow. You can do the same thing in the y-direction, too. Taking the two together gives you a summary of how, at that moment, the landscape is changing at the point (x, y).

For the full-dress version all we have to do is go up a dimension, though we lose the nice geometrical image. Still, the meaning of the Laplacian is the same: at every point, at each moment in time, it measures changes in the way heat is flowing in the x-, y- and z-directions. We have to multiply it by a because different materials allow heat to move through them in different ways, but other than that we've really got an answer to the question on the left: "How does the temperature at a chosen point change over time?" That's a rough idea of what the Heat Equation says.

Understanding how heat flows required new mathematical inventions; physics was never quite the same again.

The Heat Equation

The Wave Equation

From swimming pools to violin strings, this fundamental equation describes how waves behave.

Acceleration

Laplacian of position

$$\frac{\partial^2 u}{\partial t^2} = c^2 \nabla^2 u$$

Propagation speed of the wave

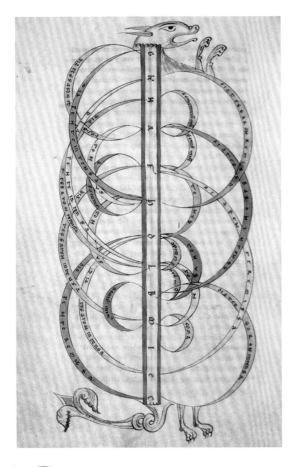

What's It About?

When a guitarist plucks a string, you might imagine that something very simple is happening: the ends of the string are fixed in place and the middle is free to move, so it must surely vibrate backwards and forwards in the obvious way (see The Damped Harmonic Oscillator, page 78). The truth, though, is much more complicated than that.

A clue to this might come from the fact that a guitar and a piano sound very different. You might say this has to do with the shapes of the instruments, and how the note is sounded (by being hit with a mallet rather than plucked), and all that's true to an extent. Yet there's something else to consider: the way their strings vibrate. For they don't simply move up and down in a nice neat fashion; slow-motion photography reveals extremely complex movements, and these have a strong influence on the sound of the instrument.

The Wave Equation gives us a much more accurate model of a vibrating string. Its richness comes from the fact that in any particular situation there are an infinite number of "basic" solutions

Boethius's diagram of the modes of a vibrating string.

A Mirror up to Nature: Science

The Wave Equation can describe waves in two- and three-dimensional media as well as the simple one-dimensional string.

to the equation, and these can be combined to produce a sort of "super-solution" in which all the possible vibrations of the string happen together. The basic solutions correspond to what physicists call "partials," which musicians call "overtones" or "harmonics."

You can get a good feel for this by finding a friend and each taking one end of a long, light rope. Stretch it out so it's not touching the ground but not taut either, and move your hands to create a steady wave, as if you were turning a skipping rope. This is one fundamental solution. Now move your hands twice as fast, and you should be able to create a new pattern, with the rope hardly moving at all at its exact center but vibrating enthusiastically between the central point and each

of your hands. The point where it isn't moving is called a "node." You may find that by moving your hands even faster you can produce other patterns of vibration, too, with more nodes along its length. Each of these is another fundamental solution to the Wave Equation.

Why Does It Matter?

The Wave Equation isn't just about musical instruments. It applies equally well to electromagnetic waves (see Maxwell's Equations, page 92) and waves in fluids (see The Navier-Stokes Equation, page 96). It works for both "standing" waves, like the ones in a guitar string, and "traveling" waves like the ripples that spread through a pond. It's just as good with the shock waves caused by volcanoes and earthquakes as it is with microwaves and x-rays. Since elementary particles can also be thought of as waves, it plays a central role in quantum mechanics, too (see The Schrödinger Wave Equation, page 104).

In technology, it makes SONAR and Synthetic Aperture RADAR possible along with more specialized methods of imaging and surveying. Among these are several techniques used by oil and gas companies mapping underground reservoirs of their favorite hydrocarbon and the ultrasound machines doctors use to survey the inner parts of us.

The problem of describing waves is historically important, too. It was explored by several of the great 18th-century mathematicians, especially d'Alembert, Euler, Lagrange and Daniel Bernoulli. All of them made good progress, but their answers seemed to contradict each other. It was only with Fourier in the early 1800s that the idea of multiple solutions, all equally valid and capable of being combined, came into proper focus (see The Fourier Transform, page 138). This idea was itself of huge importance and helped crystallize the role of abstract mathematics in physical science.

In More Detail

This version of the Wave Equation is rather general, since it works in any number of dimensions. To get into the detail of it, let's suppose our setup is the rope stretched between you and a friend, and we'll only worry about up-and-down motion. So any point along the rope can be identified by the distance from your hand — call that x. When the rope is pulled tight, all those points are in a straight line at the same height — call the height u, and call this starting height $u = 0$. Finally, we want to watch the wave waving and that requires time, which as usual we'll call t. So we're looking for a function u that takes in a distance x and a time t, and returns a height: call it $u(x, t)$, for short.

Notice that we know everything about the wave if we know the function $u(x, t)$ — that is, given a position along the rope (x) and a time (t), we can say how high that point will be at that time. From this we can reconstruct what the rope looks like at any moment, and by repeating this for different moments we could create an animation of the waving rope that's as accurate as we like. Finding this function $u(x, t)$ is what we mean by finding a "solution" to the Wave Equation.

On the left of the equation is the rate of change of u with respect to t. Fixing any point x, it tells us how that point's accelerating at that moment. On the right is the Laplacian of u (see The Heat Equation, page 80). This tells us how, if we freeze time, the height is varying close to each point. For example, suppose the rope is at the top of its vibration. If there were a tiny ant standing at the halfway point, it would say the rope looked pretty flat, just bending away very gently on both sides. If the ant were standing near your hand, though, there'd be a more obvious slope. The Laplacian tells us, very broadly speaking, how fast that slope appears to be changing as the ant looks along the rope.

The Laplacian is being multiplied by c^2; c itself is the speed at which the wave moves through the material. In the case of a standing wave, like the one in your stretched rope, you can think of it as a number made up of the tension in the string and its density. Either way, squaring it makes the units agree on both sides of the equation, which is

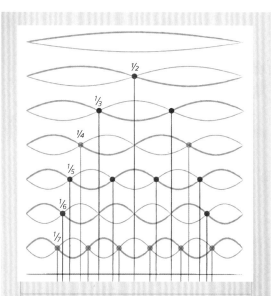

The simplest ways a guitar string can vibrate are its harmonics, which divide it into a whole number of vibrating sections. In real life it vibrates in more complicated ways that combine these possibilities.

A Mirror up to Nature: Science

terribly important (see $E = MC^2$, page 88).

Finding solutions to the Wave Equation — that is, finding functions $u(x, t)$ that make it true — was an important problem in the 18th century that motivated some extremely important mathematical research. Evidently, the different ways you can get the stretched rope to vibrate are all solutions. It turns out these can all be modeled using sine waves, which are particularly simple functions that model oscillating movement (see Trigonometry, page 14). An acoustic sine wave, incidentally, sounds very pure and simple, a bit like a flute.

Fourier's idea was that if you have some solutions to the Wave Equation you can make lots of new ones just by adding up multiples of what you already have. This produces many more complicated

Even when playing the same note, the flute and violin sound different because they represent different solutions to the wave equation.

solutions — all of them representing ways the rope could vibrate if you could only persuade it to. In most musical instruments, we find the waves are like this: they can be closely modeled by adding up several sine-wave-based solutions to the Wave Equation (see The Fourier Transform, page 138). This goes for other wave-like phenomena in nature, too. This principle — called "the principle of superposition" — is incredibly useful in many situations involving differential equations, and hence especially in many areas of physics.

A close relative of the Heat Equation, the Wave Equation is a kind of supercharged version of the Damped Harmonic Oscillator, giving a rich and precise model of periodic processes.

$$E = MC^2$$

**It's the most famous equation in all of physics.
But what does it mean, and why is C squared?**

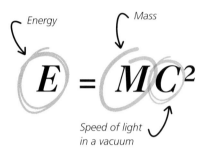

Energy

Mass

$$E = MC^2$$

Speed of light
in a vacuum

What's It About?

Einstein's equation says something astonishing: that mass and energy are really the same thing looked at in different ways. This is distressing not just scientifically but philosophically. Mass seems to deal with matter, the stuff the universe is made of. Matter is just what there is, and it seems it can't be created or destroyed, only converted into different forms; the more matter there is, the more mass.

We seem to be able to imagine a universe that has mass but no energy, in which inert lumps of stuff just lie around. But can we imagine a universe with energy and no mass? It seems not, since energy captures something about what stuff does, or could do. A boulder on a cliff-edge has potential

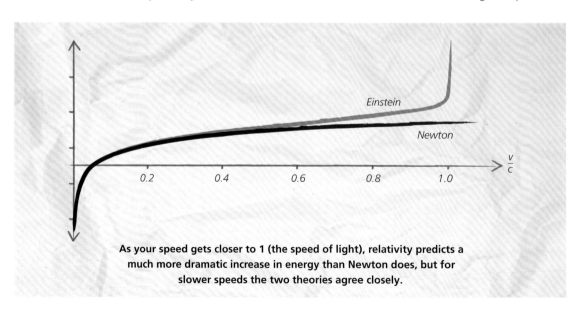

As your speed gets closer to 1 (the speed of light), relativity predicts a
much more dramatic increase in energy than Newton does, but for
slower speeds the two theories agree closely.

A Mirror up to Nature: Science

energy because gravity could cause it to fall; if it did, it would convert potential into kinetic energy while it fell, and when it reached the bottom that would get converted again into sound, shock waves, heat and the shattering of the boulder: the same energy in different forms. Could anything like that happen if there were no boulder and no ground?

So it seems that mass is the stuff that exists, and energy comes second as something that can belong to it. Einstein's equation says that this picture is all wrong. "Mass" and "energy" are just different names for the same thing — whatever that is.

Why Does It Matter?

I suppose this is the most famous equation in the world. I haven't tried this, but if you asked a random selection of people on the street to write down an equation they knew, I bet this one would come up far more than any other. For that reason alone, it matters: for science exists within society, and most of us aren't scientists, so when science captures our imagination enough that it makes an equation famous — an equation, of all things! — something significant has happened even if the science isn't important.

Well, as you probably already know, the science that gave birth to Einstein's equation was important. It changed a set of basic assumptions about how the universe works that had been around since the late 1600s. In particular, it overturned Newton's picture of space and time, which had underwritten pretty much all of physics since then, and replaced it with something much weirder: curved space, time dilation and the equivalence of matter and energy, for starters.

Although, it doesn't seem to agree with our everyday experience at all, Einstein's equation turns out to describe certain extreme situations rather well. Newtonian physics works fine as a model

The corrections Einstein's theory makes to Newtonian mechanics make a significant difference to the accuracy of the GPS system.

$$E = MC^2$$

of the universe most of the time, and that hasn't changed: when things are medium-sized and not moving too fast, that is. The trouble comes when things have enormous mass or are going at extreme speeds. Think of relativity as providing an important adjustment to the Newtonian model that's really, really tiny in most situations — so tiny it can be ignored. It can't, though, be ignored in fields like astrophysics, where it often makes the difference between a model that makes good predictions and one that makes rotten ones.

There are a few areas of technology, too, in which relativity is important. The most famous example is the Global Positioning System (GPS), a network of 24 satellites that you can use to determine your position on the Earth's surface with an amazing degree of accuracy. This requires some not-very-complicated calculations involving position, speed and time, but basing these on Newtonian physics leads to inaccuracies that would mess up the system and make it significantly less useful. Relativity provides a correction that improves the system.

The most famous application of the specific equation we're looking at here is much less happy, although Einstein often gets too much of the blame

for it. The equation says that mass and energy are just different names for the same thing. Nuclear bombs, therefore, don't convert mass into energy, which wouldn't really mean anything. What they do is convert the energy stored in certain materials into a different, much more destructive force. Einstein's equation helped scientists to understand how this process works and calculate the amount of energy released, but it's probably a mistake to think it alone made the atom bomb possible.

In More Detail

We begin with something from classical mechanics: kinetic energy, which is a measure of how hard we have to work to stop a moving object in its tracks (see Conservation of Angular Momentum, page 62). A baseball thrown by a professional pitcher has lots of kinetic energy. It has less when I throw it because I can't throw it as fast as the pitcher can. On the other hand, there's more energy in the situation if we swap the ball for a car going at the same speed because the car has more mass. That's why it would hurt more to get hit by the car than by the baseball.

In a formula, kinetic energy has the following definition:

$$K = \frac{1}{2} \, m v^2$$

where m is mass and v is velocity. Look slightly familiar? Going from this to $E = MC^2$ takes us from the Newtonian equation to the relativistic one. The star player is the basic assumption of Special Relativity: that the speed of light in a vacuum is the fastest anything in the universe can travel.

Suppose a baseball is traveling through space at close to the speed of light. That's pretty fast, but maybe you're flying alongside it, so it doesn't seem that way to you. And suppose you have lots and lots of energy to spare. You decide to give the baseball a whack to make it go faster. What stops you sending it beyond the speed of light? Einstein's theory says that the trouble is you'll have to whack it harder and harder to get smaller and smaller increases in speed as you get closer to the speed

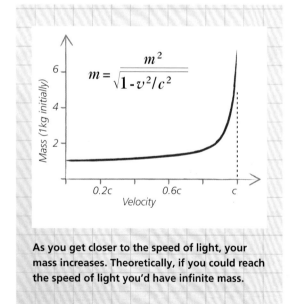

$$m = \frac{m^2}{\sqrt{1 - v^2/c^2}}$$

As you get closer to the speed of light, your mass increases. Theoretically, if you could reach the speed of light you'd have infinite mass.

A Mirror up to Nature: Science

The faster they're going, the harder the bat has to be swung to get the same acceleration.

limit. It'll be a matter of ever-diminishing returns: you put in huge amounts of energy to get a tiny bump in speed, and the next bump will cost you even more. To get the baseball past the speed of light would need an infinite amount of energy, and nobody has access to that (see Zeno's Dichotomy, page 18).

If mass and energy are essentially the same thing, putting energy into a system (for example, to make it go faster) increases its mass; this can be observed in experiments, although the increase is very tiny, since it's equal to the energy increase divided by the speed of light squared, which is a huge number. The increase in mass means that to accelerate it a little bit more means you have to put more energy in than last time, just as pushing a car takes more energy if someone's sitting in it. It's as if the only way to push the car were to put more and more people into it, which makes pushing it harder and harder.

The factor that captures this behavior turns out to be this function of velocity:

$$\gamma(v) = \frac{1}{\sqrt{1 - v^2/c^2}}$$

Notice that when $v = 0$ this factor is equal to 1, and as v gets closer and closer to c the factor gets very large — in fact, it shoots off to infinity. We can now write the Special Relativity version of the kinetic energy formula as

$$e = \gamma m c^2$$

When the object is stationary relative to the observer, this simplifies to the famous $e = mc^2$.

Energy and mass seem to be completely separate concepts: this equation's radical proposal is that they're not really so different after all.

$E = MC^2$

Maxwell's Equations

The theory of electromagnetism opened the way for a huge array of modern technologies as well as the "field theories" of physics.

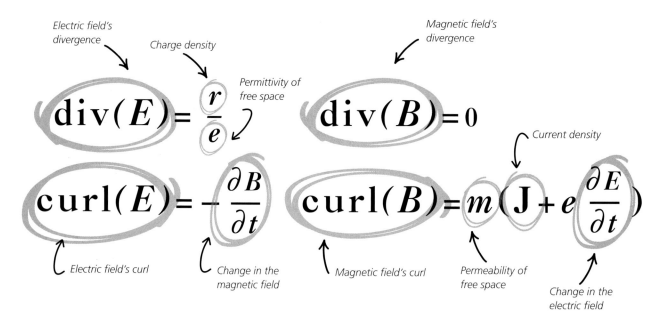

Electric field's divergence

Charge density

Permittivity of free space

$$\mathrm{div}(E) = \frac{r}{e}$$

Magnetic field's divergence

$$\mathrm{div}(B) = 0$$

Current density

$$\mathrm{curl}(E) = -\frac{\partial B}{\partial t}$$

Electric field's curl

Change in the magnetic field

$$\mathrm{curl}(B) = m\left(J + e\frac{\partial E}{\partial t}\right)$$

Magnetic field's curl

Permeability of free space

Change in the electric field

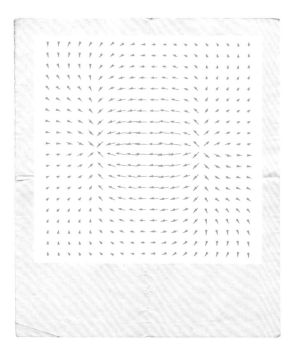

What's It About?

Magnets and electricity have probably fascinated people for as long as there have been people who noticed them. Brilliant, crackling sparks seem to appear out of nowhere from fur, quartz, rubber or amber; magnets move inanimate objects without touching them; and lightning is (if you'll pardon the pun) one of the most striking events we see in the sky. There's something magical about all these, and we can read about early encounters with them in texts from many different cultures, where they're often treated as what we would now call supernatural phenomena.

More level-headed inquiries also went on, but it wasn't until the late 18th century that these forces began to be grasped scientifically, and even then, the results were perplexing and the theories they produced incomplete. An old observation that magnets create a field — spreading out

A Mirror up to Nature: Science

their effects in the space around them — was reinvigorated by Michael Faraday's experiments in the early decades of the 19th century and gradually became more rigorous and better understood.

The crowning achievement of this development was James Clerk Maxwell's Dynamical Theory of the Electromagnetic Field in 1868, which unified electricity, magnetism and light in the same theoretical framework. Electromagnetic fields are modeled by vector fields (see The Hairy Ball Theorem, page 46) and Maxwell's four interconnected equations describe their fundamental properties.

Why Does It Matter?

It's not much of an exaggeration to say that we owe a great deal of 20th-century technology to Maxwell and his contemporaries. Without them it's hard to see how radio, television, computers, microwaves, lasers, x-rays, mobile phones, fiber optics, wireless internet and hundreds of other modern wonders could have been invented. Even if their principles had been stumbled upon in some other way, the understanding necessary to engineer them with precision would have been

missing. The electromagnetic theory is also useful in "pure" science, especially cosmology. After all, electromagnetic radiation, including light, is the main way we observe distant objects in the sky.

What's more, Maxwell's theory was a step towards an ambition that physicists are still working towards today: Grand Unification. This will hopefully result in a single theory encompassing all four fundamental forces in nature: gravity, electromagnetism and the strong and weak nuclear forces. Perhaps the final unified field theory, if we ever achieve it, will be as elegant as Maxwell's unification of the apparently different forces of electricity and magnetism. In the 1970s various steps were made in this direction, notably the unification of electromagnetism and the weak force into electroweak theory, but the reconciliation of Special Relativity with quantum mechanics (and of gravity with the other forces) remains elusive.

Iron filings in a magnetic field line themselves up along the directions of the vectors described by Maxwell's Equations.

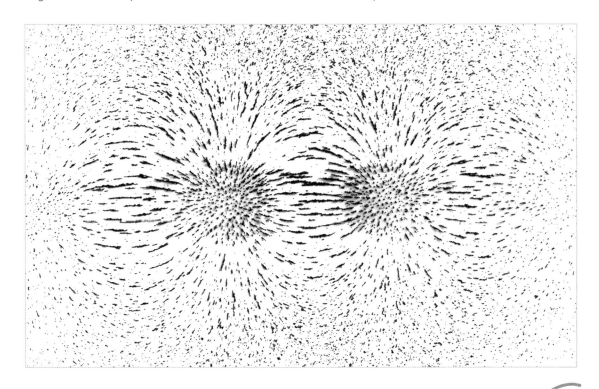

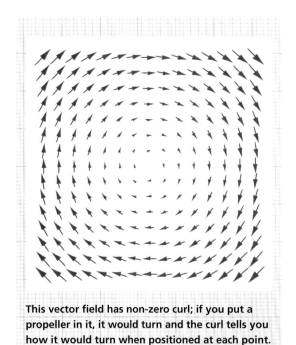

This vector field has non-zero curl; if you put a propeller in it, it would turn and the curl tells you how it would turn when positioned at each point.

In More Detail

These four equations use two ideas from vector calculus: divergence ("div") and curl. Like the Laplacian (see The Heat Equation, page 80), these describe the geometric behavior of a field close to a point, but whereas the Laplacian dealt with fields of numbers representing things like temperature or displacement (see The Wave Equation, page 84), these new operations apply to fields of vectors. Incidentally, you're looking at what are called the differential forms of Maxwell's equations; there are equivalent formulations in terms of integrals (see The Fundamental Theorem of Calculus, page 26). Though they look scary, the thing to remember is that together they aim to give us a very physical description of a phenomenon that previously seemed mysterious and ineffable. While engineers and physicists find both forms useful, we'll stick with the differential versions.

The top two equations are about the divergence of the electric and magnetic fields, respectively. Divergence at a point in a vector field says to what extent the little arrows very close to that point are

pointing away from it, on average.

Because electromagnetic fields are invisible, they're not very easy to get a mental picture of. They do, though, share some properties with fluids, so we can get the general ideas by analogy with something more familiar. Picture a swimming pool in which people are splashing around, so the water is swirling in a complicated fashion. Attach a little arrow to each water molecule showing which way it's going and how fast (indicated by the length of the arrow). Pick a point somewhere in the swimming pool and look at a very small volume of water around it. There might be lots of arrows pointing all over the place, but unless something weird's going on we expect that, overall, the same amount of water will be flowing out of that small volume as flowed in, because water molecules don't just appear out of nowhere or vanish into thin air. This means that the divergence at that point — and so at every point — is zero.

Now, let's suppose someone pops a hose into the pool and starts pumping water into it. At the point at the end of the hose, water is rushing away in all directions and not much, if any, is flowing towards it. We say this point has positive divergence, and we can measure it with a number using a spot of calculus: the faster the flow out of the hose, the bigger the number. Similarly, if, instead of pumping it in, the hose was sucking water *out* of the pool, that point would have negative divergence, with lots of arrows going towards it and few, if any, pointing away.

The first of Maxwell's Equations says that the divergence at a point in an electric field is equal to the charge density at that point divided by a constant that captures how electric charges behave in a vacuum. If there's no electric charge the charge density is zero and so is the divergence — that is, no electricity is flowing. If the charge is positive, the divergence will be positive too, so the field is flowing away from that point. As you might suspect, when the charge is negative the field flows towards it, as expressed by the resulting negative divergence.

The second equation is even simpler: magnetic fields never have any divergence. In electricity, you

A Mirror up to Nature: Science

have separate positive and negative charges and those, as we've just seen, generate the divergence of the field. In a magnet, though, you never have just a north pole or just a south pole on its own; you always have both together. This means that at any point those two cancel each other out, and the divergence disappears.

In the third and fourth equations we move from divergence to curl, which is a measure of how much the vector field is "circulating" around a point. For an intuitive picture of this, let's go back to the swimming pool and imagine inserting, at a specific point, a little propeller attached to a light rod. The flow of the water might cause the propeller to spin. We move the end of the rod around, keeping the propeller in the same spot, until we find the angle that makes it spin fastest: the curl at that point tells you two things at once: the direction the rod's pointing and the speed the propeller's going.

The third — known as Faraday's Equation — says that a circulating electrical field at a point is created by a magnetic field that's changing over time. This is the principle behind generators that use moving magnets to produce electricity, so if you're looking for one equation that essentially gave us the modern world, there it is.

Finally we come to the Ampère-Maxwell Equation, which effectively says the opposite: the curl of a magnetic field is produced by a changing electric field. This is what gives us the electric motor: again, since you're probably surrounded by those right now, it's not hard to see its importance.

Magnetism and electricity both seem to act at a distance, as if by magic. This makes them useful for all kinds of purposes.

In a breakthrough in our understanding of how the universe works, Maxwell used cutting-edge math to unify disparate phenomena about magnets and electric fields into a single, simple picture.

The Navier-Stokes Equation

The flow of fluids is described by an equation that's still not fully understood; a million-dollar prize is available for anyone who can solve it.

Acceleration

Pressure gradient

External forces

$$\rho\left(\frac{\partial v}{\partial t}+v\cdot\nabla v\right)=-\nabla P+\mathbf{div}(T)+f(x,t)$$

Viscosity

Divergence of the stress tensor

What's It About?

Newtonian mechanics gave us a pretty good model for ordinary, solid objects like rollercoaster cars, springs and cannonballs (see Newton's Second Law, page 56). The world isn't only made of those sorts of things, though: it also contains fluids, which are considerably weirder. Liquids and gases are both fluids; movements of the atmosphere and oceans are, of course, crucial parts of what makes our environment the way it is.

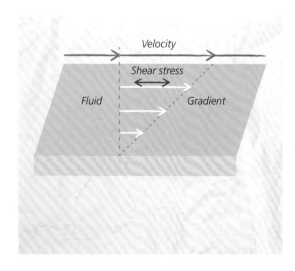

Velocity

Shear stress

Fluid

Gradient

The physical behavior of fluids is pretty complicated, so making a decent model of them isn't easy. First imagine dropping a ball onto the floor — although it's a bit fiddly, we can use Newtonian physics to predict where and how the ball will bounce around, and if we take account of enough things like friction, air resistance, bumps on the floor and so on we can make pretty good predications about it. Now imagine doing the same, not dropping a ball but emptying a bucket of water onto the floor. How does the water move through the air as it falls? Can you predict the way it will splash when it lands, or which parts of the room will get wet?

Although it moves around in a somewhat complicated way, the ball fundamentally stays ball-shaped and its behavior is, to a large extent, determined by that. The water, on the other hand, changes shape radically in its transition from the bottom of the bucket via a vertical stream to a puddle on the floor. As it changes shape its physical behavior changes too, and that makes things tricky. In fact, when so many things are changing in such complicated ways all at the same time it's surprising we can say much at all about fluid flow; that's the achievement of the Navier-Stokes Equation.

A Mirror up to Nature: Science

Why Does It Matter?

The Navier-Stokes Equation belongs to the classical physics that began in the time of Newton: not the quantum world of the very small or the relativistic worlds of the very massive or very fast-moving, but the kind of physics that describes the phenomena we meet in everyday life. In fact, it's one of the very few parts of that kind of physics that remain mathematically mysterious.

A decent model of fluid flow is useful in a huge range of situations. Industrial processes involving fluids and scientific studies of fluid phenomena like oceans and the atmosphere are obvious examples; aeronautical engineering, in particular, makes heavy use of them. In recent years, even movie visual-effects teams have got in on the act, using the Navier-Stokes model to create realistic CGI simulations.

We don't know if these equations always have "well-behaved" solutions, meaning that in some situations they might radically fail to be accurate models of fluid flow. For example, a solution might suddenly shoot off to infinity, even though we know that infinite velocity is a physical impossibility (see $E = MC^2$, page 88). Understanding the

equation better would bring us a big step closer to completing classical physics; the Clay Institute in Providence, Rhode Island, has offered a million-dollar prize for solving this mystery, but at the time of writing nobody has been able to claim it.

A solution to the Navier-Stokes Equation is a vector field that attaches a little arrow to every point in the fluid, indicating the direction and speed of the fluid's flow at that point (the speed is represented by the length of the arrow; see The Hairy Ball Theorem, page 46). If we had this solution, we'd have a complete picture of the fluid's motion at that moment in time and, if we let time move forwards, we'd see the velocity vectors changing as the fluid flowed, giving a sort of animation of the whole movement. Well, this is the point of the equation; it doesn't say much about how it's put together.

In More Detail

The best approach with a complicated beast like this is to break it down into manageable pieces. Sure, in the details there's a lot of advanced calculus and a fair bit of physics to understand, and to use it to make a realistic model of a fluid

The movement of a liquid depends, in part, on how viscous it is.

The Navier-Stokes Equation

requires some serious chops. Still, as an equation it's really not so bad once you get to know it. I should mention before we start that "the" Navier-Stokes Equation is quoted in many different forms, and sometimes as more than one equation; the way I've stated it is, I think, just about the easiest to make sense of.

On the left-hand side the thing in the brackets is technically the "material derivative" of the velocity, but we can consider it as roughly equivalent to how the fluid's velocity is changing at each point, which is pretty much the same as its acceleration. This is being multiplied by a constant number, ρ, that represents how viscous the fluid is — whether it's runny like water or gloopy like lentil soup. Acceleration is the rate of change of those velocity vectors we're trying to find. Remember that in physics we often know accelerations before we know velocities or positions. That's because we can often figure out all the forces involved, and force and acceleration are intimately linked

The Navier-Stokes Equations can be used to model gases as well as liquids.

together (see Newton's Second Law, page 56). So, the equation is going to give us a way to calculate the acceleration at each point and each moment in time, and the solution will turn this into a velocity.

Notice that the effect of ρ is to make the acceleration smaller. If this isn't completely obvious, imagine the equation is like this:

$$\rho \times [\text{acceleration}] = [\text{some stuff we know how to calculate}]$$

The stuff on the right gives us what *could* be an acceleration vector, but to find the actual acceleration we have to divide it by ρ first. This will scale the acceleration down if ρ itself is a big

A Mirror up to Nature: Science

number. This is what we expect: all other things being equal, lentil soup shouldn't accelerate as easily as water.

On the right of the equation we have three terms, each of which is a little model of something that contributed to the acceleration of the fluid at the point x. These are, in order, the pressure gradient, the stress tensor and any external forces that are involved.

The pressure gradient is worked out by standing at our chosen point and pointing in the direction that the pressure is decreasing fastest: that's a vector, which is $-\nabla P$. We know that pressure is intimately related to the state the fluid is in — particularly, its temperature and volume (see The Ideal Gas Law, page 66). Experience, if nothing else, tells us that any difference in pressure causes movement from highest to lowest pressure: this is why you keep getting yanked out into space when you open that airlock. Really, this is just the same as saying that the pressure gradient has an effect of acceleration, and that's why it's been invited to this party.

Something similar is happening with div(T), the second term. Here T is a close relative of the stress tensor (see the Cauchy Stress Tensor, page 122), an object that collects together information about the forces acting on a small cube around our point of interest. It's there to take account of movements within the fluid that are causing parts of it to tend to shear relative to other parts

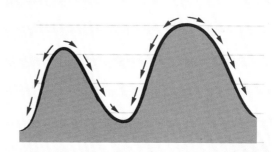

The graph shows pressure as height. Liquids tend to flow away from high-pressure areas into lower-pressure ones — that is, they flow down the pressure gradient.

nearby — compression, the other aspect of the stress tensor, is already accounted for by ∇P. When we take its divergence we get something similar to the pressure gradient: a vector indicating how the shearing stresses are changing close to each point.

The last term is a bit of a catch-all: f simply gathers up any extra forces you might want to apply to the fluid to get an accurate model. You almost always want f to include the force of gravity, otherwise your model will predict that, when you tip your cup over, the coffee will hover about in a sort of blob instead of making a nasty puddle on the tablecloth.

The Wave Equation is only part of the picture when it comes to how fluids move in the real world. The Navier-Stokes Equation gives a much more widely applicable one.

The Navier-Stokes Equation

The Lotka-Volterra Equations

A simple but powerful model for how populations grow and decline.

Change in wildebeest population →

Wildebeest birthrate

How much lions eat

$$\frac{dx}{dt} = rx - axy$$

Number of wildebeest

Number of lions

Change in lion population

$$\frac{dy}{dt} = -mx + bxy$$

Lion death rate

Lion birthrate

What's It About?

A herd of wildebeest lives on a large island with plenty of whatever wildebeest like to eat on hand. Left to themselves they will reproduce freely until they run out of space and food. The wildebeest are not, however, left to themselves; they share the island with a group of lions. When there are plenty of wildebeest around, the lions thrive and increase in numbers.

As they do, though, they eat more and more wildebeest, whose population declines. The island can no longer support so many lions, so their numbers begin to drop, too. Now the wildebeest have a better chance of surviving and reproducing, so their numbers go up. As more wildebeest become available, life gets easier again for the lions … and so the cycle continues. The Lotka-Volterra Equations were designed to capture this apparently simple interaction.

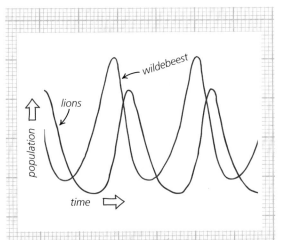

The two curves show how the lion and wildebeest populations change over time according to the Lotka-Volterra model.

A Mirror up to Nature: Science

Why Does It Matter?

The Lokta-Volterra Equations were originally designed for modeling biological predator-prey systems and they are still used that way, but they often need adjustment because the assumptions they make aren't always terribly realistic. For example, if there are no lions at all, the wildebeest population grows exponentially forever, because when $y = 0$ the situation simplifies to

$$\frac{dx}{dt} = rx$$

From a little bit of calculus it follows that

$$x = e^{rt} + c$$

which says that the population is growing exponentially! This means that if we begin with a single mated pair of wildebeest and no lions, and suppose they produce two offspring per generation, within 100 generations there'll be more wildebeest on the island than there are atoms in the observable universe, a conclusion most biologists would find somewhat implausible.

We do sometimes see this kind of growth in real life, if only for short periods of time. In the 1860s, European colonists in Australia systematically released rabbits into the wild to provide a food source. The rabbits reproduced so fast that local carnivores couldn't make a dent in their numbers, which grew more or less as the Lotka-Volterra Equations predict, devastating crops in the process. Within a decade the policy of introducing rabbits into Australia was replaced by frantic attempts to get rid of them. This has left behind a dramatic monument to exponential population growth: the famous "rabbit-proof fence," actually three connected fences extending for more than 2,000 miles (3,000 km), designed to prevent the spread of the "plague" from east to west. The rabbits' initial period of unrestricted growth, though short-lived, was so explosive that Australians are still struggling to control their numbers today, a century and a half later.

Instead of populations of competing biological

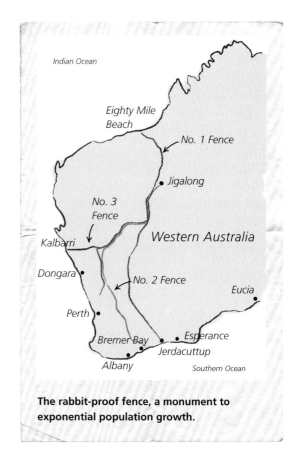

The rabbit-proof fence, a monument to exponential population growth.

species, the equations can be adapted to describe the prospects of agents competing for any scarce resource, so they've been used in economics, sociology and even finance. They've also been employed to describe aspects of resource management, repressive government, neural networks and game theory. As with any mathematical model of complex phenomena, though, these applications are always approximations, and they're only as good as the assumptions behind them.

In More Detail

The Lotka-Volterra Equations are one of those things whose moment had to come: although human beings had been watching predator-prey interactions for thousands of years, the equations were invented quite independently in the early

1920s by two separate researchers, both of whom were, in different ways, outsiders.

Alfred J. Lotka spent almost his whole career outside the world of academic research, effectively working as an amateur. He saw the world as a huge, interconnected system in which physics, chemistry and biology all interact, governed by the same basic principle of the exchange of energy. It was this vision that led him to derive these equations from a simpler model for chemical reactions. Chemistry, physics, biology: fundamentally, it was all the same to him.

Vito Volterra had been born into the Jewish ghetto of Ancona, Italy, but managed to work his way through the educational system and become a distinguished research scientist. It was only at the age of 65 that he investigated the fluctuating populations of sharks and stingrays in the nets of Adriatic fishermen and ended up proposing a model exactly the same as Lotka's. Five years later he refused to pledge allegiance to Mussolini and was stripped of all his academic positions. He spent his last decade wandering Europe, still writing books about mathematics.

A key feature of this model for both its discoverers is that it's periodic; that is, it repeats the same pattern over and over again. You can see a suggestion of this in the graph on page 100, but there's a better way to visualize it using a visual gadget called "phase space."

There are two varying quantities: x, the number of wildebeest, and y, the number of lions. We can set up x- and y-axes representing them so that each point on the plane represents a possible state the two populations can be in. Now we attach an arrow to each point that indicates how the model predicts that a population in that state will change — a sort of "flow" from one state to another.

For example, if there are a lot of lions and not many wildebeest, we're somewhere around the upper left of the image. In this situation, many lions will go hungry, so the arrow will point sharply downwards. When there are many lions and many wildebeest (upper right), the arrow indicates that the lions will still be increasing a bit but the wildebeest numbers will be dramatically declining.

Near the bottom there are few lions, meaning wildebeest numbers can increase quite freely, indicated by the arrows pointing to the right. Once there are "enough" wildebeest, the arrows start pointing upwards, showing the number of lions starting to grow.

Once you have a nice set of arrows drawn in — this is a bit tiresome, but a computer can do it in the wink of an eye — you can "see" the behavior of the system very easily. Imagine that the phase-space diagram is the surface of a lake seen from above, and the arrows represent currents. If you drop a stick into the water, the point where it lands represents the initial populations of lions and wildebeest; the arrows show how the stick will be pushed around by the currents over time. As the solid lines — called "attractors" — show, this model predicts that the stick will move around and around in closed cycles forever.

The picture also shows us that if we drop the stick in just the right spot, it won't move at all: this is an "equilibrium point" for the system. In other words, the model predicts that we could find a perfect balance of lion and wildebeest numbers that would be stable: for every wildebeest eaten,

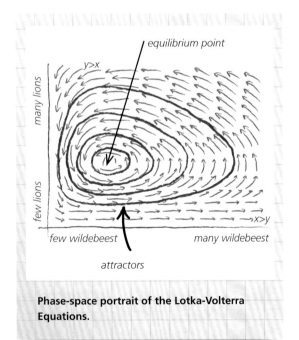

Phase-space portrait of the Lotka-Volterra Equations.

A Mirror up to Nature: Science

another would be born. Although it would be rare to find perfect equilibrium in nature, it seems possible to get close to it, and the stick dropped near the equilibrium point will cycle around it very tightly, with little boom and bust in either population.

What really complicates things, though, is that we very rarely have just two populations in competition with nothing else going on. The Lotka-Volterra Equations can be extended to cope with this, but there's a cost: chaos. With just three species, even very tiny variations in the starting conditions produce wildly different long-term results. The attractors in phase space can become "strange attractors" that have a fractal structure, one of the signature features of chaotic systems. The first strange attractors in Lotka-Volterra systems were discovered and classified at the end of the 1970s. This was part of the birth of chaos theory, the study of systems that are governed by simple rules but behave in extremely complicated ways. How these models evolve over time for even modest numbers of species is still not well understood today.

The Lotka-Volterra Equations were one of the places chaos theory got started. With two species everything's fine, but adding a third leads to wildly unstable behavior.

$+ z =$ CHAOS!

It's quite a simple model of the relationship between predator and prey but its evolution over time is surprisingly complicated and can even be chaotic.

The Schrödinger Wave Equation

Everyone's heard of Schrödinger's cat; its big brother, the Wave Equation, explains what's going on behind the scenes.

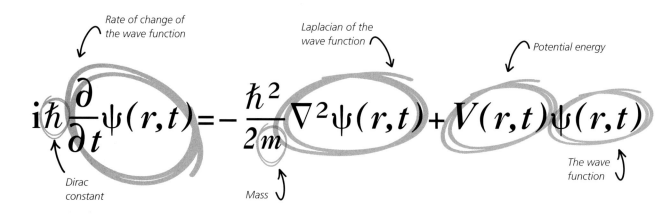

Rate of change of the wave function

Laplacian of the wave function

Potential energy

$$i\hbar\frac{\partial}{\partial t}\psi(r,t) = -\frac{\hbar^2}{2m}\nabla^2\psi(r,t) + V(r,t)\psi(r,t)$$

Dirac constant

Mass

The wave function

What's It About?

In classical mechanics we use the relationship between force and acceleration to find an equation of motion for an object that, with a little extra observational information, can tell us where the object is, how fast it's moving and in what direction (see Newton's Second Law, page 56). This usually involves setting up an equation with some forces that we can hopefully know, plus a term involving acceleration, the second derivative of position. "Solving" the equation means calculating the acceleration and then using calculus to find the velocity and position (see The Fundamental Theorem of Calculus, page 26).

In the quantum world things are done more or less the same way, but what changes is that "position" and "velocity" don't quite work the way we expect them to. This is because at this level a particle can also be considered as a wave; that is, something that's spread out in space instead of being concentrated at a point. Instead of solving an equation of motion, then, we solve a wave equation: the Schrödinger Wave Equation, to be precise.

Why Does It Matter?

This is one of a handful of equations in quantum mechanics that have real foundational importance. The particle-wave duality is philosophically mysterious, but the resulting predictions are highly accurate and paint a strange but powerful picture of how the universe works at very small scales. Sadly, the weirdness of the subject has led some over-enthusiastic writers to draw fanciful conclusions from it that the science can't really support.

Physicists, though, are still working on unifying this picture with Einstein's, which deals with the very large and fast (see $E = MC^2$, page 88). If they're successful they will have achieved a "theory of everything" that will be an incredible intellectual achievement and may well bring new scientific and technological breakthroughs along with it.

In more obviously practical areas of life,

A Mirror up to Nature: Science

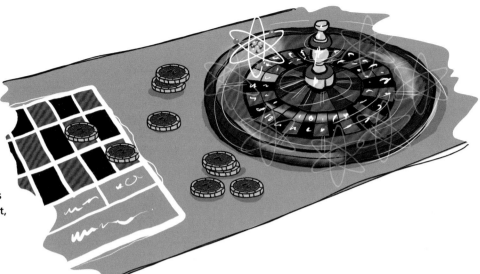

Does God play dice? The Wave Equation is neutral on the subject, but how it should be interpreted is still a controversial topic.

quantum mechanics is the science behind semiconductors and lasers, without which we'd have none of the miniaturized electronics that now surround us. Other, newer applications are emerging, too: quantum computers, superconductors, nanotechnology and new and exotic materials. These are already entering our lives and might well revolutionize them before the end of this century.

In More Detail

This is one of those scary-looking equations with a lot of moving parts, so let's break it down into

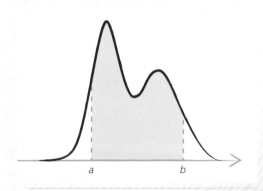

The probability of finding the particle between points *a* and *b* equals the size of the shaded area.

its main bits. The star of the show is the symbol $\psi(r, t)$, which is the wave function itself: usually, this is the function we're trying to find, just like the *x* in $3x + 2 = 8$. This is a more complicated equation that's more difficult to solve, but the same basic idea applies. If we can find out what function $\psi(r, t)$ is, then we know how the particle wave is behaving at the point *r* and at time *t*.

On the left-hand side of the equation we're differentiating with respect to time; in other words, we have the rate of change of the function as time moves forward. This is conceptually something like a velocity, but it isn't the velocity of a little billiard-ball-like particle. After all, we're no longer doing classical physics and our particles are now also waves. So, instead of a little ball moving with a certain velocity, we have a wave function that evolves over time.

This "velocity," if you're still willing to call it that, is being multiplied by the constant number $i\hbar$. The letter i represents $\sqrt{(-1)}$ and indicates that we're using the complex numbers to represent two-dimensional space (see Euler's Identity, page 40); this is just a mathematical convenience and needn't detain us. The other symbol, pronounced "h-bar," is more significant. It comes from the Planck-Einstein Relation, $e = h\nu$, which relates the energy of a particle (*e*) to its frequency as a wave (*v*) via a number called Planck's Constant (*h*).

The symbol $\hbar$ just represents $h/2\pi$, another bit of tidying-up, and is often called the Dirac Constant.

On the right-hand side of the equation we have two objects being added together. The first is the Laplacian of the wave function; if you're willing to see the first derivative as a velocity, you can see this Laplacian as a sort of acceleration that captures the way the velocity is changing. This is again being multiplied by a constant factor, this time involving the mass of the particle as well as $\hbar$. Multiplying a mass-like term by an acceleration-like term might vaguely remind you of $F = ma$ (see Newton's Second Law, page 56), which is the distant ancestor of this equation.

The second object is there to represent the external effect of any electromagnetic field the particle might find itself in (see Maxwell's Equations, page 92). This is natural enough: if you picked some particle or other to look at, it wouldn't usually exist in perfect isolation from everything else, and if it did it wouldn't necessarily be all that interesting. The function $V(r, t)$ represents the strength of that field at the point r and at time t; the time component is needed because these fields typically change over time just as the wave function does.

So that's the Schrödinger Wave Equation: the rate of change of ψ (times a constant) is the acceleration of ψ (times a different constant) plus the effect of the electromagnetic field the particle's in. One of its earliest successes was in calculating the possible energy levels an atom of hydrogen can have: these can't vary continuously, but are separated out into "quanta," like rungs on a ladder. The equation (combined with some linear algebra) is able to confirm their values with great accuracy.

This is all very well, but what does the wave function really represent in the physical world?

At the 1927 Solvay conference in Copenhagen, Niels Bohr and Albert Einstein wrangled over the probabilistic interpretation of quantum theory in the company of some of the most eminent physicists of the time.

A Mirror up to Nature: Science

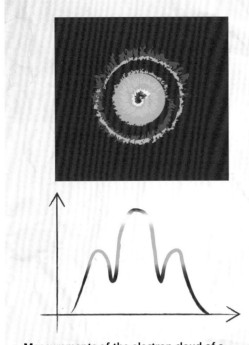

Measurements of the electron cloud of a hydrogen atom (top) correspond closely to the predictions of the Wave Equation (above).

for it there at time t. When we do observe the particle, we "collapse the wave function," turning it into a simple statement that the particle is at that location; but the wave function itself tells us what the chances were of making that observation at that moment. This is daring because it feels as if the particle doesn't really exist at the point x at all. It's more that the particle is a spread-out field of probabilities and that our observation makes one of them happen. Seen in this way, the wave function is like the distribution of probabilities at the roulette table and an observation consists of actually spinning the wheel and seeing where the ball ends up. It's not that the result was always 35 (say) and that by making the observation we discovered this fact: rather, the observation made it true.

This is certainly a troubling idea. Einstein, famously, preferred the interpretation that the particle is at the place where it's observed, and the observation just confirms it. On this view the element of probability in the Wave Equation reflects a defect in the physical theory, not a real indeterminacy in the universe: "God does not play dice." A competing perspective, though, is offered by the so-called Copenhagen Interpretation of quantum mechanics, which was devised in the 1920s. It says we should bite the bullet and agree that the universe is fundamentally probabilistic: in a sense our observations actually create the things they're measuring. Part of the attraction of this is that it helps us understand certain predictions made by quantum mechanics that seem to require faster-than-light travel, which relativity suggests is impossible (see $E = MC^2$, page 88). Bell's Theorem of 1964 shows that this means quantum mechanics can't be given a classical interpretation, no matter how clever it is. The quantum world really seems to be as strange as those pioneering physicists suggested.

That's a much harder question to answer. After all, we started off being interested in a particle — something like an electron, say — and we've ended up with a wave function that describes a phenomenon that's spread across all the available space (that is, the function has a value at every point). What does the Wave Equation mean?

One daring suggestion is to treat it statistically. Specifically, the number

$$|\psi(r,t)|^2$$

is interpreted as the probability of finding the particle in a region very close to r if we look

Quantum mechanics seems to turn the deterministic universe of classical physics into one governed by chance and probability, at least on one reading of this equation.

The Schrodinger Wave Equation

$$|O(\bar{1}, z, a, b)| = 2$$

$$\varphi(\sigma_1 t)\, \varphi(\sigma_2 t) = \varphi(\sqrt{\sigma_1^2 + \sigma_2^2}\, t) \qquad \sum_{k=1}^{r} \int_{b \in \nu} \left(\int_0^b \psi_k^*(\tau)\, d\tau \right) dt$$

$$\omega) = \frac{\sum_{k=1}^{r} P_k^\alpha \log_2 \frac{1}{P_k}}{\sum_{k=1}^{r} P_k^\alpha} \qquad C_{ik}\bar{\sigma}_h^2 = \lambda_i\, C_{ik} \qquad \eta_1 = \sum_{k=1}^{n} a_k \xi_k$$

$$y = \phi(x) = \frac{1}{\sqrt{2\pi}} \int_{-\infty}^{+\infty} e^{-\frac{t^2}{2}}\, dt \qquad S(\alpha, \bar{1}$$

$$= A_n \cup \Pi A_u \qquad W_k = \binom{n}{k} p^k (1-p)^{n-k} \qquad P(\eta < y \mid \xi = x) = \sup_{y' < y, y'}$$

$$|n| = \frac{n!}{2} \quad \left| \int_{|x| > A} f(x) \log_2 \frac{1}{f(x)}\, dx \right| < \varepsilon \qquad g^{-1} \cdot g = e$$

$$d\, G_n(x) \geq \frac{1}{2} \sum_{h \to \infty}^{+\infty} e^{-\frac{h^2 \pi^2}{\lambda^2}} = H(\lambda) \qquad \prod_{k \leq b} ; \bigcup_{i=1}^{n-1} M_i ; \bigcap_{n=0}^{\infty} X_u$$

$$f_{n-1}(t) = \int_0^1 f_n(u)\, f_1(t-u)\, du = \frac{\lambda^{n+1} t^n e^{-\lambda t}}{n!} \qquad \lim_{t \to 0} (\varepsilon(t)) = 0$$

$$\log \varphi(t) = i\gamma t - c\, |t|^\alpha \left[1 + i\beta \frac{t}{|t|} \omega(t, \alpha) \right] \quad B(v) = \sum_{k=1}^{r} \psi^*(b_k v) \qquad C_{iv} =$$

$$e^{-\frac{u^2}{2}}\, du = F(x) \left(\frac{1}{\sqrt{2\pi}} \right)^{-1} \quad |\psi_\xi(t)| = \left| \int_{-\infty}^{\infty} e^{itx}\, d\bar{f}(x) \right| \leq \int_{-\alpha}^{\alpha} e^{-vx}\, dF(x) =$$

$$\Pi_m = \Pi_r \Pi_{m-r}$$

$$|X \cup \Psi| = |X| + |\Psi| - |X \cap \Psi| \qquad \lim_{n \to \infty} \frac{1}{\sqrt{n}} k_n \left(\frac{x}{\sqrt{n}} \right) = \frac{1}{\sqrt{2\pi}} e^{-\frac{x^2}{2}}$$

$$f : X \to X \cap W$$

$$(A) = \int_A \chi(\omega)\, dP \qquad l'(\alpha) = -\log 2 \left(\frac{\sum_{k=1}^{r} P_k^\alpha \log_2^2 \frac{1}{P_k}}{\sum_{k=1}^{r} P_k^\alpha} - \left(\frac{\sum_{k=1}^{r} P_k^\alpha \log_2 \frac{1}{P_k}}{\sum_{k=1}^{r} P_k^\alpha} \right. \right.$$

$$\left(e^{-x} \sqrt{\frac{1-q}{nq}} - 1 \right) = -x \sqrt{\frac{q(1-q)}{n}} + O\left(\frac{1}{n}\right) \qquad \prod_{k=1}^{r} \left[g_k \left(\frac{t}{\sqrt{N_0}} \right) \right]^{N_0 \alpha_k} = e^{-\frac{c t}{2}}$$

$$\liminf_{} \int_{}^{+\infty} f_N(x)^\alpha\, dx \geq \int_{}^{+\infty} f(x)^\alpha\, dx$$

What Have You Done For Me Lately?

TECHNOLOGY

The Mercator Projection

**What choices are involved in making a flat map of the world?
What are the consequences and is there a "best" choice?**

Vertical position

$$v(a,b) = a$$

*Coordinates of the
point on the globe*

$$h(a,b) = \log\left(\tan\left(\frac{b}{2} + 45°\right)\right)$$

Horizontal position

What's It About?

Making flat maps isn't a big problem when the area you're charting out is fairly small, because the Earth's curvature doesn't cause major problems: it's such a small part of the whole globe that it's pretty close to being flat. What's more, small inaccuracies usually aren't a big deal if you're just trying to find your way from one town to the next.

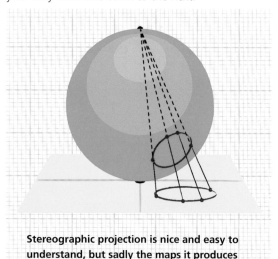

Stereographic projection is nice and easy to understand, but sadly the maps it produces are badly distorted.

With the advent of transatlantic travel in the 1500s, though, that changed. Cartographers began to make maps of the whole world, and that involved unavoidable distortions. A map is flat, after all, but the mean curvature of the globe isn't zero (see Curvature, page 30). You can't go from one to the other without giving something up: either distances and areas get distorted or angles and locations do. In other words, either the relative sizes of things are wrong or they aren't where they appear to be.

Why Does It Matter?

Mapmakers had to think carefully about the compromises they chose to make when flattening out the globe. There are many solutions to this problem, but Mercator's from 1569 is still one of the most popular. In fact it's one of a family of ways to map points on a sphere onto a flat plane; one of its relatives, the stereographic projection (see left), is important in many different areas of mathematics.

Sixteenth-century sailors used maps of the world to navigate across oceans. Small inaccuracies could send a ship many miles off course before it made land, which was very bad news if you were low on

What Have You Done For Me Lately: Technology

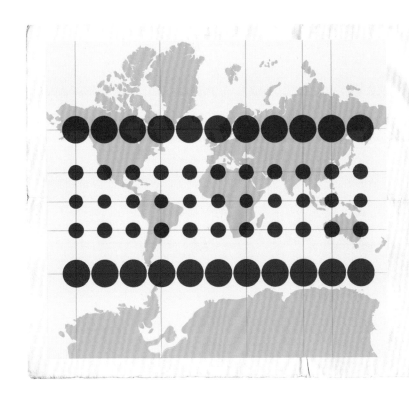

Tissot's indicatrices are circles designed to show how the Mercator Projection distorts the sizes of objects. On the globe all the circles would be the same size, but in the projection the ones near the poles are much bigger than those at the Equator.

provisions. As a result, finding a good projection from the sphere to the flat page was an important challenge. What's more, it was gradually discovered that different projections had different properties, which made them advantageous in different situations. There's no perfect choice: cartographers need multiple ways to do it, and to understand the pros and cons of each.

In More Detail

Conceptually, the Mercator Projection works in a fairly straightforward way. You can imagine it most easily by constructing, in your head, a rather fancy sort of lamp. The equations defining the projection, which were discovered later by the 17th-century mathematician Henry Bond, just provide a mathematical way to relate each point on the globe to a point on the map without resorting to this contraption, which as you'll see would be a bit inconvenient on board a tea clipper.

Suppose you have a glass globe with an accurate representation of the world drawn on it.

In the center of the globe put a small, bright bulb. Next, put this whole setup inside a tall, cylindrical lampshade made of thin paper that touches the glass globe along its Equator. Turn on the lamp and what happens?

The idea is that the paper cylinder will be our map — notice that, although it seems curved, we can unroll it into a flat surface without any distortion. Every ray of light leaving the bulb goes in a straight line. It hits the globe at a certain point — suppose it's in the middle of an ocean, so the light passes straight through the clear glass. It then carries on and eventually hits the lampshade. That spot on the paper cylinder is the spot on the map that represents the spot on the globe that the ray of light originally passed through.

On the other hand, when a ray of light hits a part of the globe that's the edge of a country, the light will be absorbed and won't hit the lampshade. As a result, on the lampshade we'll see the edges of the countries marked out. What you're looking at is the Mercator Projection. To create a flat map we simply cut the cylinder vertically and unroll it; usually this cut is made through the Pacific Ocean. In reality, though, we use the two equations at the top of page 110 to translate angles of latitude and longitude (see Spherical Trigonometry, page 114) into the x- and y-coordinates of the flat map.

You might be worrying about rays of light that go up through the Arctic and come straight out of the top of the shade, and you'd be right to do so. I wasn't kidding when I said it had to be a tall

The Mercator projection transfers the Earth's surface onto a cylinder using rays from the center.

cylinder. In fact if you want the whole globe, you'll need it to be infinitely tall, and even then you'll be missing the two isolated points at the exact North and South Poles. In reality, though, this isn't a huge practical problem. As you move away from the Equator the distortion of the projection gets worse and worse, making it more or less useless in the area of the poles, so if you're sailing in that region you'll need a different map anyway. With this in mind, a moderate-sized cylinder will give you a decent map, missing out two circular regions of the Arctic and Antarctic.

The Mercator Projection is now understood to be one of a family called the Lambert Conical Projections; in fact, it's an extreme example. The whole family can be understood by just modifying our imaginary lampshade, leaving the rest of the setup exactly as it was.

Instead of a cylindrical shade, use a conical one. The cone can be steep (like a witch's hat) or shallow (like the Asian sedge hat) and in each case when you cut from the point of the cone straight down to the edge and unroll it you'll get a different kind of projection. With a conical projection, notice that at the pointy end of the cone you don't lose anything the way you do when rays of light escape at the top of the cylinder — the North Pole, in fact, is projected onto the point itself. The price you pay for this is that you've had to widen out the hole at the bottom; this means you lose a bigger part of the southern hemisphere centered on the South Pole.

Now, imagine the cone getting flatter and flatter; eventually it'll just be a flat sheet of paper lying on top of the globe, and what you see will be a form of "stereographic" projection. Rays of light from the central bulb that go through points on the northern hemisphere get mapped onto the paper; those that pass through the Equator will be parallel to the page, so they never meet it. Nor do those rays going through any point south of the Equator, which are heading away from the page as soon as they leave the bulb.

With a little imagination you can start with the cylinder; picture it transforming into a very tall, steep cone with the point very, very high up;

What Have You Done For Me Lately: Technology

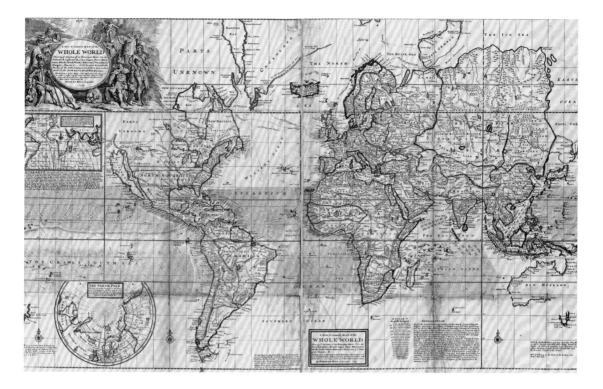

picture the point moving slowly downwards, so the cone becomes ever less steep; and finally picture that point touching the North Pole, at which moment the page is flat. This continuous series of setups produces the Lambert Conical Projections.

Projections have many applications outside mapmaking. The stereographic projections, in particular, come up everywhere from perspective drawing to the theory of complex numbers (see Euler's Identity, page 40), although often the "bulb" is placed on the South Pole — with a bit of careful definition, and an infinite sheet of paper,

Mercator's map from 1569. Although distorted, the compromises he chose preserved the information most helpful to navigators. That made it an important tool in this age of long-distance trade.

this means we can "map" the whole sphere except for the precise point at the South Pole itself. The problem with that point is unavoidable; mapping it using these projections is a bit like an eye trying to look at itself.

Turning an oblate spheroid like the Earth into a flat map is hard. The Mercator Projection was one of the earliest and most successful attempts to negotiate the compromises required.

Spherical Trigonometry

Triangles on the surface of the Earth don't behave the way they do on a blackboard; understanding that makes intercontinental flights and GPS possible.

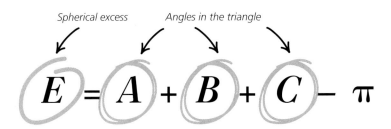

Spherical excess   *Angles in the triangle*

$$E = A + B + C - \pi$$

What's It About?

A bear walks a mile south, a mile east and a mile north, ending up back where he started. What color is the bear? Perhaps you've heard this old chestnut before; if not, see if you can solve the riddle before reading on. So, what's the answer? Well, the only way it's possible to do what the question described is by starting at the North Pole, and we all know what color bears around there tend to be. This happens because the Earth isn't

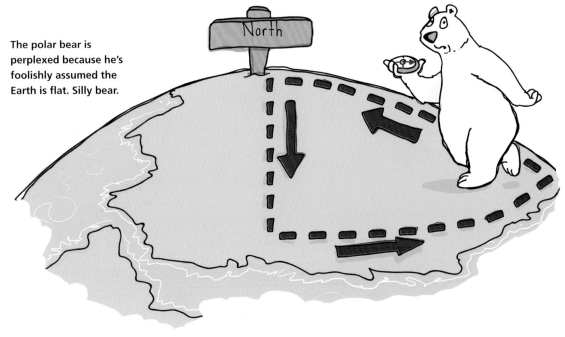

The polar bear is perplexed because he's foolishly assumed the Earth is flat. Silly bear.

North

This armillary sphere, used for astronomical calculations, divides the celestial globe into eight triangular regions, each with a spherical excess of 90°.

flat like the blackboard in a geometry classroom: it's curved, and that curve can make a big difference.

On a flat surface like the blackboard, any triangle's internal angles add up to exactly 180°, or half a turn. On the surface of a sphere, though, they never do. You can investigate this by drawing triangles on something curved, though it needn't be a perfect sphere — a balloon will do. If you pick the right spot, it's possible to make a triangle with

two right angles *and* another one, like the one the bear walked around, even though that makes a lot more than 180° altogether.

This is an indication that the geometry we learned at school is going to lead us astray if we try to apply it to triangles on curved surfaces like the Earth. Sure, in small cases like laying out triangular buildings or fields we can level off the ground to make it flat, but when much bigger triangles come into play that's no longer an option.

Why Does It Matter?

It's very useful to be able to find our way precisely around two-dimensional spaces: the surface of the Earth is one such space, and the screen of a computer or television is another. Many spatial and geometric problems can be reduced to problems about triangles (see Pythagoras's Theorem, page 10, and Trigonometry, page 14), so having a good handle on how triangles behave is definitely a step in the right direction. Yet the two situations just described are quite different: on the screen, the triangles live in a flat space, while we all know that the surface of the Earth is curved into a sphere.

This makes a big difference to our measurements. If we use what we know about triangles on flat spaces like the blackboard or a page from a notebook to make calculations about distances and angles on the Earth's surface, the results will be wrong. On a small scale they work OK as an approximation, but when we're looking at large triangles — those that describe intercontinental flights, for example — the results will be off by a rather dramatic amount.

We first met such large triangles when we began making long-distance voyages. Navigation using a flat map, and flat triangles, can turn into a disaster (see The Mercator Projection, page 110), and it's in this context that spherical trigonometry — literally the study of triangles on spheres — was born.

Since then the field has been useful in astronomy and geography, including those recent applications of satellites for finding our way around and making beautiful images of parts of the world we'll probably never get to visit in person.

In More Detail

Trigonometry (see page 14) is about triangles, which are closed shapes with three straight sides. What, though, counts as a straight line on the surface of a sphere? If you pick two points on a ball and try to put a ruler between them to draw the straight line, you'll find you can't, because the ruler juts out of the sphere instead of following its surface.

Instead, imagine you're an ant walking around on the sphere and want to get from A to B as directly as possible. It turns out that you'll walk along what's called a "great circle," which is a circle whose center is also the center of the sphere, so that it cuts the sphere exactly in half. If we assumed the Earth was a perfect sphere (which it isn't), the Equator and the Greenwich Meridian would be great circles but the Tropic of Capricorn and the Arctic Circle wouldn't. As an aside, notice

A round trip between London, Moscow and Cape Town forms a spherical triangle if routes between the cities are direct.

that on a sphere there are no parallel straight lines: every pair of straight lines must meet at two points, which are on exactly opposite sides of the globe from each other.

Incidentally, this is why nautical miles are a bit different from ordinary miles. The latter are units of measurement based on a flat surface, which is fine for relatively short distances. If you're traveling along a great circle at sea, though, it's more natural to measure how far you've gone around it in degrees. A nautical mile is the distance you travel if you go 1/60 of a degree around a great circle; 1/60 of a degree is called a "minute" and if you go this far in an hour you're traveling at a speed of 1 knot. Spherical trigonometry measures the edges of its triangles in this way, too. This makes it all about angles, and the separate idea of "length," so familiar to us from flat geometry, vanishes almost completely.

Anyway, it makes sense to define a spherical triangle as a three-sided figure whose sides are all parts of great circles. For example, if you travel in an exactly straight line from London to Moscow, then similarly from Moscow to Cape Town and from there directly back to London, your path forms a spherical triangle. Spherical trigonometry relates the positions of these three points on the globe to the lengths of the three journeys and the angles you have to turn through at each point. This is obviously very important in navigation.

It also crops up in astronomy. When you were a child, perhaps you looked up at the night sky and imagined it as a solid dome, high above you, in which the stars were embedded. In fact this model — extended to imagine it as a sphere enclosing the Earth on all sides — is still useful when making practical observations and calculations, even though cosmologists assure us it's not at all right. In this setting, too, spherical trigonometry, now on the inside of a sphere, is extremely useful. Not only is it used today, its importance in early astronomy means we should consider it just as ancient and venerable as the so-called ordinary trigonometry we learn at school.

Our equation calculates the *spherical excess* of a given triangle. That's simply how much more

What Have You Done For Me Lately: Technology

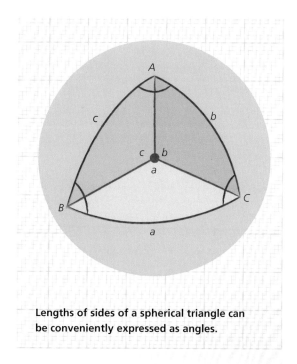

Lengths of sides of a spherical triangle can be conveniently expressed as angles.

Surfaces of negative curvature, on the other hand, have triangles with an *angle defect*; that is, their angles sum to less than 180°. Such situations are much less common in our everyday world, though.

The fact that geometry is different on the sphere from the flat plane has been used in various decorative settings. A mundane example is the soccer ball, which is often covered entirely by a mixture of pentagons and hexagons. If you look closely you'll see all the pentagons and hexagons are regular, meaning their sides and angles are all the same — and of course they fit together nicely to cover the surface of the ball without any gaps. If you try to cover a flat piece of paper with regular hexagons and pentagons in this way, you'll soon find it can't be done. That's because on the flat piece of paper, the angles at a point must add up to 360 exactly, but the spherical excess on the soccer ball gives us more elbow room, enabling us to fit those corners together.

This fact is also exploited on a larger scale in the decorations found on the domes of some buildings, especially those in the Islamic world where abstract geometric patterns were preferred to representational images. The problem of "tiling" a dome with polygons was already in an advanced state by the 10th century, when it was codified by Abul Wafa al-Buzjani. Since then, extremely elaborate patterns have been constructed on domes and similar structures that would be impossible on a flat surface — all made possible by the spherical excess.

than 180° its angles add up to, though we use radians for neatness (see Euler's Identity, page 40). This is always a positive number, because spheres have positive curvature (see Curvature, page 30) — yes, even the insides of spheres. After all, we can imagine our sphere is made of glass, so that we can look at a triangle drawn on it from either the inside or the outside. The geometry of a shape shouldn't be changed by how you look at it!

When you work on a sphere instead of a flat plane, geometry changes rather dramatically and some of the things you learned at school stop being entirely true.

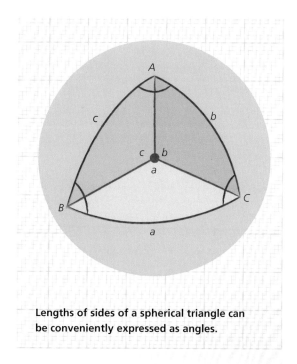

Spherical Trigonometry

The Cross-Ratio

Perspective distorts distance, direction and even proportions; the cross-ratio, though, always stays the same.

Ratio of lengths in real life

Ratio of lengths in any image

$$\frac{AC}{CB} \bigg/ \frac{AD}{DB} = \frac{AC'}{CB'} \bigg/ \frac{AD'}{DB'}$$

What's It About?

It's true that we live in a three-dimensional world, where the laws of ordinary school geometry are in force. It's also true that we live on the surface of a sphere, so that sometimes we have to use a different kind of geometry (see Spherical Trigonometry, page 114). Yet we also live in a world of projections: flat, 2D representations of 3D worlds, whether real or imaginary. Every time we capture a photograph or a video, we're flattening out the space we live in, and our world can seem filled with images others have made in the form of entertainment and advertising. The relationship between two-dimensional images and the three-dimensional spaces and objects they represent is part of the remit of projective geometry, and the cross-ratio is at its heart.

We know from experience that a photograph can distort many things. For example, if you took a picture of this book you could frame the shot in such a way that it seemed tall or short, and you could make its corners, which in the real world are right angles, seem sharper or blunter. So we can't

Looking at an object from different perspectives distorts lengths, angles and the relationships between them.

What Have You Done For Me Lately: Technology

In Caravaggio's painting *The Lute Player*, the lute looks right even though the perspective view has distorted its proportions. The secret lies in not distorting the cross-ratio.

trust the lengths and angles we see in a picture to be the same lengths and angles in the real world. In fact, given a real-world triangle, we can, at least in theory, find a viewpoint from which it looks like any other triangle we choose. So much for all that geometry we learned at school.

What's worse, we can't even trust the proportions of things to stay the same. You might have noticed this when you've seen a picture of someone reaching towards the camera, which can cause their arm and hand to seem weirdly out of scale with the rest of their body. You might start to wonder whether there's anything at all that stays the same in projections. The answer is yes: the cross-ratio.

Why Does It Matter?

We've probably all learned to be a bit sceptical about that old saying, "The camera never lies."

Still, though, we use photographs as evidence in scientific and legal settings, among others. The police use automated cameras to catch speeding motorists, CCTV images to construct descriptions of suspects and crime-scene photographs that identify the locations of people and objects. The forensic analysis of 2D images often needs the help of projective geometry, and the cross-ratio is especially useful because it has a relationship to reality that many other qualities of the photograph lack.

These techniques extend to many technological applications involving optics, including the frontiers of computer vision and automatic

image interpretation. The cross-ratio has helped researchers working on difficult problems such as face recognition and inferring the geometry of a 3D space from a 2D image. Projective geometry has wider applications in various sciences, including astronomy.

There's a more fun set of applications, too, in creating 2D images of imaginary 3D worlds. Essentially, the same set of procedures is used in both CGI on the big screen and video games at home, both of which explicitly use projective geometry to pass from a model of a 3D object constructed entirely within the computer to a representation from a particular viewpoint on the screen in front of us. Developers even talk about that viewpoint as a "camera" and imagine it moving around a solid object or space. In fact, there is no camera: the geometry is doing all the work in the abstract.

It's tough to create a perspective drawing when extreme foreshortening is involved, as when an arm reaches out towards the viewer.

In More Detail

It's a common claim in how-to-draw books that an adult human male is about eight times as tall as the height of his head and that his belly-button is about 4.5 heads up from the ground. Let's calculate the cross-ratio of this specimen. First we'll label him starting from the ground up: the soles of his feet are at A, 0 heads above the ground. His belly-button is at B, 4.5 heads up. His chin is at C, 7 heads above the ground, and the crown of his head is at D, 8 heads up.

To find the cross-ratio we need four lengths: AC, CB, AD and DB. The first of these is the

What Have You Done For Me Lately: Technology

distance from ground to chin, so $AC = 7$. The next is from chin to belly-button; the distance is 2.5 but this is measuring downwards, against the positive height direction, so we say $CB = -2.5$. The whole height is $AD = 8$. Finally, DB is the downward measurement from crown to belly-button, so $DB = -3.5$. Altogether that gives us

$$\frac{AC}{CB} / \frac{AD}{DB} = \frac{7}{-2.5} / \frac{8}{-3.5}$$

$$\approx -2.8 / -2.29 \approx 1.22$$

Now, wherever the camera is, and whatever angle this man is at, and even whatever units of measurement we use, this number will always come out the same. The right-hand side of the equation says that the cross-ratio of the distorted lengths (AB', for example, rather than AB) will be the same as that of the originals. You can take a picture that distorts the man's proportions to some extent — for example, from a camera high above him — but you can't distort his cross-ratio. In fact, distorting cross-ratios makes the image look really weird, as if seen in a funfair mirror.

Proportions have nothing to do with scale or units of measurement: if my head is one-third the size of my body (which it isn't), then it will be one-third the size of my body whether we're measuring in centimeters, inches or Egyptian Royal Cubits, and an action figure of me (don't laugh) would have to have exactly the same proportions as the actual me. The cross-ratio is a ratio of ratios or a "proportion of proportions," so it has these same properties, too. One place you can see projections very easily is shadows. When you stand under a streetlamp your shadow is elongated, but its proportion of proportions will be the same as yours.

To get an intuitive picture of a projection, imagine a three-dimensional space with a special point, called the "origin," which you can think of as the viewpoint where your eye is positioned. For you to see a point in 3D space, a ray of light must come from that point and hit your eye. Since light rays travel in straight lines, we think of projective points — that is, things we see as points — as infinitely long straight lines going through the origin.

Now, a projection can be thought of as a flat plane, like a sheet of paper, that sits somewhere in the 3D space. This is the image. For any 3D object you want to make a picture of, just draw a straight line from every point in the object to the viewpoint; where the line passes through the sheet of paper, make a dot. This is precisely how perspective drawing works and you can see it illustrated very nicely by some of Albrecht Dürer's 16th-century pictures of drawing machines.

Think of something like a pencil, hanging in space, put the plane between the pencil and the viewpoint and think of some of those rays of light going between the two and making the picture on the plane. Now, a projective transformation in this setup would be a matter of moving the plane: literally shifting it around or rotating it. Imagine what happens to the picture of the pencil as the plane turns in space; the line representing the pencil might get longer or shorter, even though the length of the pencil itself isn't changing. This is what we mean when we say that projective transformations don't preserve length.

If we replace the pencil with our idealized human figure, the problem remains the same, but now we can measure the cross-ratio, not just on the actual figure but on any projective image of it. When we do, we get a number that never changes. So, if someone has a CCTV picture they claim is of you, its cross-ratio ought to be the same as yours, or any other picture of you. If not, it can't be you.

We see two-dimensional projections of three-dimensional objects all the time; this obscure ratio is the closest we can get to something that stays the same despite all the distortions.

The Cross-Ratio

The Cauchy Stress Tensor

Structural engineers use this exotic object to capture the behavior of real objects when forces act on them.

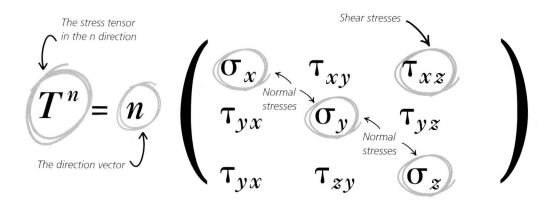

The stress tensor in the n direction

The direction vector

Shear stresses

Normal stresses

Normal stresses

$$T^n = n \begin{pmatrix} \sigma_x & \tau_{xy} & \tau_{xz} \\ \tau_{yx} & \sigma_y & \tau_{yz} \\ \tau_{yx} & \tau_{zy} & \sigma_z \end{pmatrix}$$

Calculations with the stress tensor can give us confidence that a daring structure like this building in Marseille, France, is actually very safe.

What's It About?

What holds solid objects together is the complicated network of forces between their atoms or molecules; when you put a mug down on a table, that's what enables the table top to resist the mug's weight without shattering. If, instead of a mug, you put a truck on the table, those forces are no longer a match for the external forces at work and the table does indeed fall to pieces.

Now imagine a situation somewhere between these extremes: say, for example, it's a lightweight table and I've chosen to take a chance and stand on it to change a light bulb. An engineer might well look at the piece of furniture groaning under its burden and see stress at work: internal forces threatening to pull the table apart.

Structural objects such as the joists that bear loads in buildings are subject to stresses too, which engineers must understand precisely to make sure they're strong enough. It's not enough to have a qualitative estimate — "That looks like it'll hold" — so we'd better have a more precise definition.

In More Detail

The Cauchy Stress Tensor provides that precise definition. A "tensor" is a mathematical object that helps us collect information together in a way that has particularly convenient features; the details aren't important here. What matters is that the Cauchy Stress Tensor is a package that contains all the different kinds of stress we might find at a given point in an object and allows us to calculate the net effect of the stress at that point. So, like a vector field (see The Hairy Ball Theorem, page 46), the Stress Tensor can be different at each point in an object and can vary over time.

Stresses in the Earth's crust have created many large-scale geographical features, including the Great Rift Valley in East Africa.

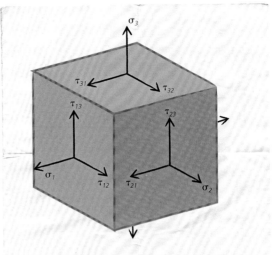

The nine components of the stress tensor working on a small cube of material.

To calculate its components — σ_x, σ_{xy} and so on — at a point in the object, we imagine the point is at the center of a tiny cube. Each component is stress acting on a different set of faces of the cube. For example, σ_x might be stress that's tending to push the whole cube up (through its bottom face) or down (through its top face). When I'm standing on the table, points in the table top under my feet definitely have this component present. Then σ_y might be stress tending to push the object from one side to the other, while σ_z would be the remaining possibility: stress going from front to back.

That accounts for stresses that might be acting directly through the faces of the little cube, but there are more. Imagine a stress that acted along two opposite faces in different directions, pushing one while pulling the other. This can happen, for example, if a beam is firmly fixed to a wall at one end and bears a weight at the other. The weight isn't causing the beam to move downwards; it's causing it to shear downwards at one end while the other end stays still. These are the other components of the stress tensor, the ones that look like τ_{xy}.

This picture of the stress in a material is incredibly detailed yet pretty easy to work with. Engineers are used to working with tensor fields and we know lots about how they behave mathematically. It turns a potentially very complicated problem about the forces between billions of molecules into something elegant and workable. Such a detailed model of stress makes possible many of the wonders of modern engineering and architecture.

Stress in an object has many components, depending on whether the object is being squeezed, pulled or sheared. A tensor is the perfect way to pack all this together into a useful whole.

The Cauchy Stress Tensor

The Tsiolkovsky Rocket Equation

The equation that launched the Space Age, not to mention the use of rocketry in warfare.

$$\underset{\substack{\text{Maximum}\\ \text{acceleration}}}{\Delta v} = \underset{\substack{\text{Rocket}\\ \text{efficiency}}}{v_e} \ln \frac{\overset{\substack{\text{Starting mass}\\ \text{(including fuel)}}}{m_i}}{\underset{\text{Final mass}}{m_f}}$$

What's It About?

In English "rocket science" is, like "brain surgery," a byword for something requiring extraordinary cleverness. We surely haven't lost our amazement at brain surgery, but perhaps rockets seem rather mundane today. You set light to one, it shoots up, it comes down again: what's the big deal? Well, it's a bit different if you're going to launch a rocket with people on board whom you'd like to get back in one piece, or if you're trying to hit a far-off target with it.

It's especially hard if you want to launch a rocket into space. The trouble is that you need enough acceleration to overcome the Earth's gravitational pull, and the force you need to generate is proportional to the rocket's mass. But to generate a lot of acceleration, you need to burn a lot of fuel quickly, and adding more fuel adds more mass to the rocket. It seems to be a paradox: to get more force, we need to add more mass, but the more mass we add the more extra force we have to generate.

In More Detail

In his 1865 novel *From the Earth to the Moon*, Jules Verne imagines the trip being accomplished by firing a vessel from an enormous piece of artillery; in 1902 Georges Méliès' famous film *A Trip to the Moon* used the same idea. Sadly such a cannon would have to be impractically vast to create the necessary acceleration, and if the craft contained human beings they'd most likely be squashed by the forces exerted on them. Although in Verne and Méliès' day a cannon was a familiar

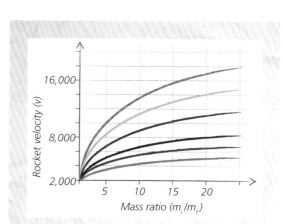

Several predictions of the rocket equation for rockets of different efficiencies (v_e). The most efficient (orange) gives the greatest velocity as the fuel burns.

What Have You Done For Me Lately: Technology

The V2 rocket was created with slave labor under appalling conditions that killed considerably more than the rockets themselves did.

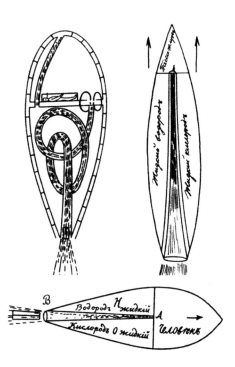

Tsiolkovsky's own designs may look sketchy, but as well as representing practical progress they inspired a new view of the future: gleaming, aerodynamic and rocket-propelled.

and well-tested way to launch projectiles through the air, it was never going to work for human space travel.

Rockets had been researched and experimented with in a military context since at least the 1810s, but not with much success. Just a year after Méliès' film, however, Konstantin Tsiolkovsky published his equation in a Russian scientific journal. At the time it made little impact, but after the 1917 revolution his work increased in prominence and influenced much of the Soviet space programme. As well as rockets, Tsiolkovsky designed airlocks, life-support systems and a spinning space station that created artificial gravity. Alongside the space race, rocket science also contributed to an arms race, most famously in the form of Wernher von Braun's V2, around 3,000 of which were launched against Allied cities during the last year of World War II.

The rocket equation tells us how much acceleration we can generate given the weight of the rocket, the weight of the fuel and the efficiency of the engine. In particular, it says that adding more fuel gives only a logarithmic increase in acceleration (see Logarithms, page 36), which means building rockets that can hold more and more fuel gives ever-diminishing returns. Like some of the others in this book, then, this equation summarizes a whole set of information in a single, neat bundle. Like others, too, it's a relatively simple model: it ignores factors like air resistance that can be quite significant. Still, all the orbiting technology we use today — from the Hubble Space Telescope and the International Space Station to the GPS system and thousands of specialized satellites — can only get into place thanks to Tsiolkovsky's equation.

Understanding the relationship between fuel, mass and acceleration is a crucial step if you want to build a rocket that actually works.

The Tsiolkovsky Rocket Equation

De Morgan's Laws

These fundamental logical laws form the basis of all computing.

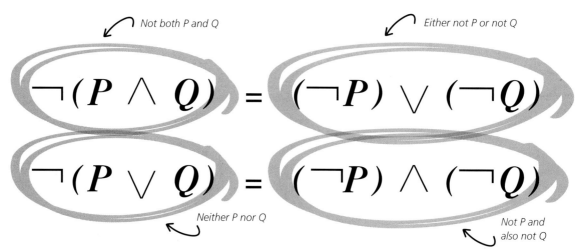

Not both P and Q

Either not P or not Q

$$\neg (P \wedge Q) = (\neg P) \vee (\neg Q)$$

$$\neg (P \vee Q) = (\neg P) \wedge (\neg Q)$$

Neither P nor Q

Not P and also not Q

What's It About?

At the heart of every computer is a central processing unit, and at the heart of that device is an array of tiny switches that can carry out logical calculations. This reliance on logic rather than arithmetic is a large part of what separates the modern programable computer from the calculating machines built in the 18th century, each of which tended to be able to do only a very limited range of different things. A modern

computer can be used to write a novel, edit a film, do your accounts or simulate complex physical systems — this kind of versatility comes from the versatility of logic.

De Morgan's Laws are among the simplest and most useful facts about logic, and when paraphrased into English they're intuitively quite straightforward. The first law says that if "It's a bird and it can fly" is false then either it isn't a bird or it can't fly (or neither). The second says that if "Either

Is it a bird, and can it fly? De Morgan's Laws enable us to reinterpret the statements about the penguin and hippo in simpler terms.

$P \wedge Q$

$\neg (P \wedge Q)$

$\neg (P \vee Q)$

What Have You Done For Me Lately: Technology

In early computers, vacuum tubes were used to create logic gates.

it's a bird or it can fly" is false, then it must not be a bird *and* it must be flightless. Computer science may, after all, be easier than you thought. Each law says that one way of saying things using "and" is exactly the same as another expression that uses "or," and gives us an explicit way to convert between the two.

Why Does It Matter?

At first glance this kind of logic might look pretty unimpressive. The two laws seem to express things that are completely obvious. Obvious to us, that is: what's impressive is that we managed to encode this obviousness in such a way that it can be etched into a tiny piece of silicon millions, sometimes billions, of times over to produce the technological wonders and distractions that now surround us, including medical devices, consumer gadgets, industrial tools and weapons systems. The encoding that made that possible is formal logic, which seems to be able to teach common-sense human reasoning to a rock (which is all, in the end, a silicon chip is).

Gottfried Leibniz had thought of building thinking-machines back in the 18th century but the formalization of logic had to be carried out before a practical, all-purpose computer could be constructed. This project took a long time.

It may be said to have started with Aristotle and was picked up again and developed in the Middle Ages in both the Arabic and European traditions. Today, the field is still home to many unanswered questions and philosophical controversies, but the huge progress made in the 19th century — the time when de Morgan derived his laws — was surely decisive. Without these developments it's hard to imagine the invention of the computer as we know it.

Towards the middle of the 19th century, Charles Babbage and Ada Lovelace made practical progress towards the construction of a programable "analytical engine." World War II saw a resurgence

De Morgan's Laws

of interest in Babbage's work, with codebreakers in Britain making important innovations and engineers at IBM creating the "Mark I," which John von Neumann used to simulate nuclear detonations as part of the Manhatten Project. After the war, Alan Turing laid much of the conceptual groundwork for modern computer science, with formal logic as its cornerstone.

In More Detail

The laws as stated belong to what's known as propositional logic, in which each capital letter stands for some statement that might be either true or false — a "proposition." In fact an important assumption is that every statement *must* be either true or false, never both and never neither, even though for a given statement we might not know which it is. In a computer we can use the presence of an electrical current at a certain point to mean that the proposition that

point represents is true, and the absence of current to mean that it's false. The logical "not" symbol,¬, simply flips the state from true to false or false to true. Imagine walking into a room wearing a blindfold, so you have no idea whether the light is on or off. Flicking the switch does what ¬ does: turns it on if it's off, and off if it's on.

We can then build up more complicated propositions by combining the basic ones using the so-called logical connectives "∨" and "∧." You can think of these as junction boxes — sometimes called "gates" — that two wires go into, with just one wire coming out. The first sends out a charge so long as it gets any charge from either of its incoming wires: in ordinary language we call it "or," though we have to remember that it includes "both." Logicians enjoy answering questions like "Is your child a boy or a girl?" with "Yes," which they take to be endearing.

The second connective only sends out a charge

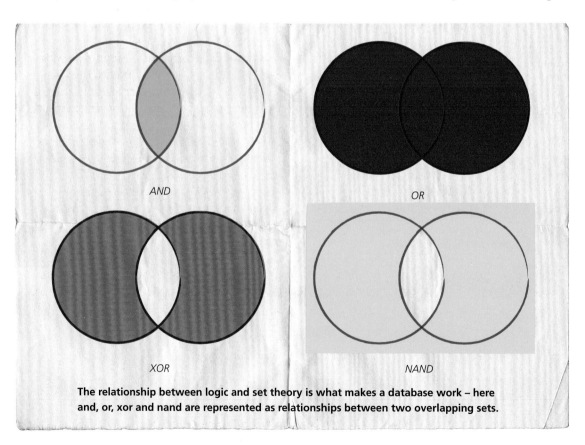

AND

OR

XOR

NAND

The relationship between logic and set theory is what makes a database work – here and, or, xor and nand are represented as relationships between two overlapping sets.

What Have You Done For Me Lately: Technology

if *both* its incoming wires have charges, so logicians and normal people alike are inclined to call it "and." The equality in de Morgan's Laws means that, regardless of the truth or falsehood of *P* and *Q*, the more complex propositions represented by the two sides of the equation must always match: they're true together or false together.

If we have lots of copies of these logic gates we can build all kinds of complicated structures, and essentially that's what most silicon chips are: collections of useful, reconfigurable tools, each of which is made out of a number of logic gates, like a sort of construction kit. Computers do arithmetic, incidentally, by representing numbers in binary form, so that the number 37 becomes 100101. Each digit is either 1 or 0, which can be thought of as "true" or "false" and fed into an array of logic gates. Even simple operations like adding small numbers require a lot of gates set up in a precise way, but the fact that it can be done — even though logic gates don't know anything about numbers or arithmetic — suggests just how flexible this approach is. De Morgan's Laws can help to simplify these complicated logical setups, either for the sake of theoretical analysis or for the practical end of reducing the amount of space each one takes up on a chip.

Although the different gates are quite convenient for human beings, computers don't actually need all those concepts. We can create "not," "and" and "or" gates using a single gate called "nand," which is short for "not and" — that is, *P* nand *Q* is logically equivalent to

$$\neg(P \wedge Q)$$

You may like to check for yourself that we can get ¬*P* from "*P* nand *P*," which is true when *P* is false and false when *P* is true. Similarly, "*P* and *Q*"

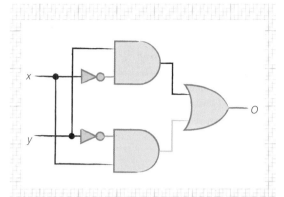

And (rounded), or (pointy) and not (small, triangular) gates making a simple circuit. What values of *X* and *Y* result in a true value at *O*?

can be expressed as "(*P* nand *Q*) nand (*P* nand *Q*)" and "*P* or *Q*" as "(*P* nand *P*) nand (*Q* nand *Q*)." These aren't as friendly for human beings to read, though their symmetry gives a hint to the debt they owe to de Morgan's Laws.

Another way to express de Morgan's Laws is in terms of set theory, which replaces propositions and logical connectives with collections of things and their unions, intersections and complements. This is important as the language of pure mathematics, but computers have given it many practical applications. One very clear example is the relational database, which is expressed and programed in set-theoretical terms. Not unrelatedly, set theory also lies at the heart of search algorithms. Indeed, you may know that some internet search engines allow you to use logical expressions that they convert into set-theoretic form in the background. De Morgan's Laws are as relevant in these domains as they are at the level of pure logic.

And, Or, Not: small words every child learns that seem to capture very basic ideas. De Morgan's Laws tie them together in a neat and symmetrical package.

Error-Correcting Codes

From the telegraph, through the Mariner mission to Mars to digital media and communications, without these codes we'd be lost in a sea of noise.

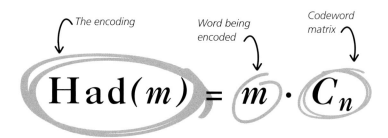

The encoding

Word being encoded

Codeword matrix

$$\mathbf{Had}(m) = m \cdot C_n$$

What's It About?

For as long as there have been complex societies, messages have had to be transmitted over long distances. For most of that time such transmissions were rather uncertain. The message had to be carried by people (often several people) and in many places travel could be dangerous: boats sank, caravans were attacked by bandits, messengers were bribed and so on. But when it did arrive at its destination, at least a letter could reasonably be expected to be the same as it was when it was originally written.

Mechanical forms of communication changed all that, though, even in simple cases like using a light to pass messages between ships. Poor visibility and mistakes made by either the sender or the receiver could introduce "noise" into the message, causing it to arrive more or less garbled. In the case of a sentence of ordinary language, human beings could often make a good guess at what the intended message had been. What, though, about a message consisting of a string of numbers? And what if at each end we had not human beings but machines, or ultimately computers that could send and receive these messages and act on them immediately?

Sometimes a message gets damaged in transit; if possible, we'd still like to be able to interpret it.

What Have You Done For Me Lately: Technology

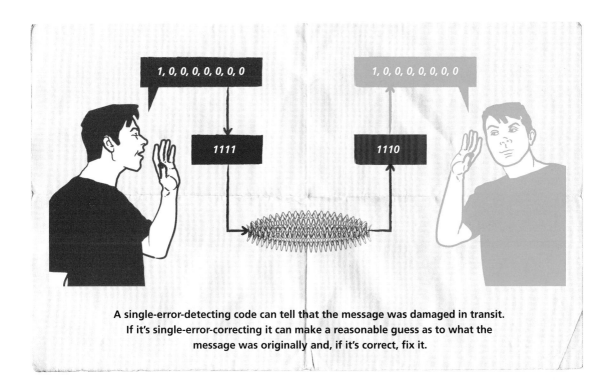

A single-error-detecting code can tell that the message was damaged in transit. If it's single-error-correcting it can make a reasonable guess as to what the message was originally and, if it's correct, fix it.

Why Does It Matter?

The simplest model of communication involves a message, a sender, a receiver and a channel — such as the postal system or a fiber-optic cable — through which the message is sent. In the real world almost all channels have "noise": physical features that tend to interfere with the message and transform it as it travels.

No error-correcting scheme is perfect. In the worst case, problems between sender and receiver could obliterate the entire message and if that happens no amount of cleverness is going to recover it. Still, it would be nice to come up with some way to cope with sending a signal through a noisy channel so that it could be reconstructed at the other end with reasonable confidence that it was right.

An obvious way to proceed if you get a garbled message is to ask the sender to send it again. The trouble is, you may not realize that it's garbled. Another approach, then, is to send every message twice — chances are that something that was messed up on the first one will be okay on the

second and vice versa. The trouble is, if you do get a difference you may not be able to tell which version's right, and it's still perfectly possible that both will be garbled.

What's more, this strategy already doubles the length of the message. That's a problem if the communication channel's bandwidth is limited, so sending a message even more times, while it improves your chances of being able to reconstruct the original perfectly, isn't really a viable option. Finally, there's the consideration of efficiency: sometimes we can recover information from horribly noisy sources, but only by expending a huge amount of effort. We'd like to be able to correct errors quickly, easily and with as little fuss as possible.

These problems only got both more severe and more commonplace as communications technology grew in the 19th and 20th centuries. Electronic communications must be accomplished quickly and cheaply, so sending the same message multiple times is much less than ideal. Over the last decade the sheer quantity of data being moved around has

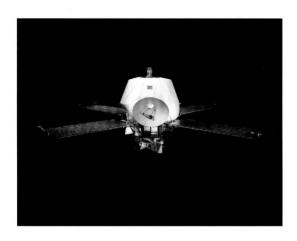

Mariner 9 used a Hadamard code to transmit detailed images of the surface of Mars.

a 1 in one particular place: for example, (0, 1, 0, 0, 0, 0, 0, 0) might be blue, (0, 0, 0, 0, 1, 0, 0, 0) might be green, and so on. This is what the *m* represents in the equation.

The first thing to do is choose a suitable Hadamard matrix, which is a special square arrangement of the numbers 0 and 1. Since we have eight letters, we'll choose the 4 × 4 Hadamard matrix (you'll see why in a moment):

$$H_4 = \begin{pmatrix} 1 & 1 & 1 & 1 \\ 1 & 0 & 1 & 0 \\ 1 & 1 & 0 & 0 \\ 1 & 0 & 0 & 1 \end{pmatrix}$$

We now stack two copies of this matrix on top of each other, but in the lower one we flip all the 0s to 1s and all the 1s to 0s:

$$C_4 = \begin{pmatrix} 1 & 1 & 1 & 1 \\ 1 & 0 & 1 & 0 \\ 1 & 1 & 0 & 0 \\ 1 & 0 & 0 & 1 \\ 0 & 0 & 0 & 0 \\ 0 & 1 & 0 & 1 \\ 0 & 0 & 1 & 1 \\ 0 & 1 & 1 & 0 \end{pmatrix}$$

To encode each color we multiply its representation by the matrix C_4. This is defined in such a way that it gives us a unique row from the matrix for each color — the second row, (1, 0, 1, 0)

grown beyond anyone's imagining, and we expect it always to arrive without a glitch, in the wink of an eye, at vanishingly small cost. That we're rarely disappointed is an extraordinary achievement; most of the time it's invisible precisely because it works so well.

In More Detail

On November 14, 1971, the unmanned Mariner 9 spacecraft entered Mars orbit and started taking photographs. It encoded the visual information as a stream of binary data and beamed it back to Earth. The messages crossed millions of miles during which they were contaminated by a huge amount of noise, yet the quality of the resulting images captured the public's imagination. To achieve this quality, Mariner 9 used an error-correcting code.

So, how does this work? Well, suppose the message we want to send is a 10 × 10 pixel digital image in which each pixel can be one of eight colors. We can send it as a stream of 100 numbers, knowing the person at the other end will be able to decode this into 10 rows of 10 pixels each and see the image. We'd like to transmit this over a noisy channel using a Hadamard code similar to the one used by Mariner 9. Initially, we'll represent each color as a list of zeroes with

What Have You Done For Me Lately: Technology

for blue, the third for red and so on. The crucial thing to notice is that any two rows are different from each other by at least two numbers. This means that to confuse, say, green with magenta there would have to be two errors, flipping the third and fourth numbers over.

Now, suppose we receive the pixel (1, 0, 0, 0). This doesn't exist as a row in C_4, so it can't be decoded right away. This pixel could have started out any color, and up to four of its numbers could be wrong. Even if we assume there was only one error, the code can't tell whether the original message was (1, 1, 0, 0) with an error in the second place, (1, 0, 1, 0) with an error in the third or (1, 0, 0, 1) with an error in the fourth.

So this code can detect a single error in a transmitted word, but can't correct it for us automatically. Still, error detection on its own is useful. No error-correcting scheme can ever be perfect; all we can do is make it more likely that we'll make the right interpretation at the receiving end. If we receive a valid code-word as a result of a series of coincidental errors, the only way we can hope to discover the problem is from the context, which is quite easy with some types of message but more or less impossible with others.

This scheme can be extended by using a bigger Hadamard matrix, although it would be impractical to show all the details here. The result is that when a limited number of errors interfere with a transmitted pixel, we can guarantee that the error can be detected and corrected. For example, suppose that, instead of differing by at least two places, the code words differed by at least four places. Then, in the case of a single error, there

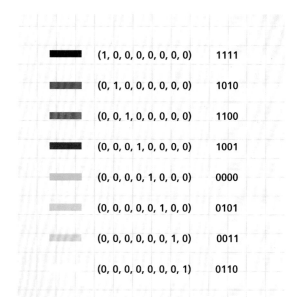

	(1, 0, 0, 0, 0, 0, 0, 0)	1111
	(0, 1, 0, 0, 0, 0, 0, 0)	1010
	(0, 0, 1, 0, 0, 0, 0, 0)	1100
	(0, 0, 0, 1, 0, 0, 0, 0)	1001
	(0, 0, 0, 0, 1, 0, 0, 0)	0000
	(0, 0, 0, 0, 0, 1, 0, 0)	0101
	(0, 0, 0, 0, 0, 0, 1, 0)	0011
	(0, 0, 0, 0, 0, 0, 0, 1)	0110

Eight colors, their initial coding as a vector and the result of multiplying the vector by C_4 to give the Hadamard code.

will always be one code word that's closer to the received transmission than any of the others are. If we guess there was only one error, we can correct it by using the closest correct word.

The payoff for this is that our messages get longer and longer. What's more, we can never be certain that we've detected all the errors or corrected them, as it were, correctly. All we can do is make it very likely, given what we know about the amount of noise our communications usually suffer.

Encoding your message in the right way can enable the person receiving it to tell whether it's been tampered with and, sometimes, even to correct mistakes introduced in transmission.

Information Theory

**A fundamental equation that's a cornerstone of
modern computer science.**

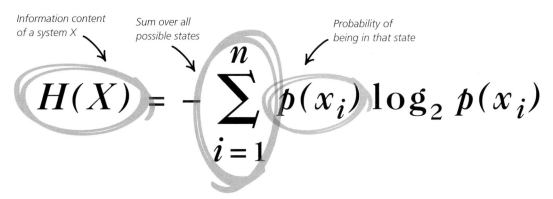

*Information content
of a system X*

*Sum over all
possible states*

*Probability of
being in that state*

$$H(X) = -\sum_{i=1}^{n} p(x_i) \log_2 p(x_i)$$

What's It About?

The most obvious way to store a digital image is simply to list the color of each pixel, going row by row and column by column (see Error-Correcting Codes, page 130). In this format a 32 × 32 pixel image would seem to contain 1,024 separate pieces of information, one for each pixel. Yet, perhaps surprisingly, modern graphics formats can compress this image to a smaller size without loss of quality.

Consider a traditional icon of a house, familiar from many web browsers:

```
00011000
00111100
01111110
11111111
01111110
01100110
01100110
01100110
```

where we're using 0 to represent white and 1 for black. The fourth row consists of eight black pixels — do we really need to say "black" eight times here? Can't we just say "eight black," perhaps using

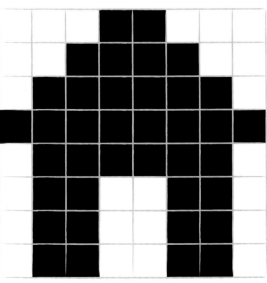

When stored as a bitmap, this simple image requires 64 bits of information: one for each pixel. But when space is at a premium, we can do better.

two pieces of data instead of eight? Looking at the bottom three rows might give us an even better idea. Instead of repeating 24 individual pieces of information, why not say "01100110, three times"?

What Have You Done For Me Lately: Technology

This is the fundamental idea behind the Lempel-Ziv-Welch (LZW) compression algorithm, which is used in standard computer-image formats like GIF and PNG. It's used in non-image formats like ZIP and PDF, too. It seems to achieve a miracle: it communicates the same picture using less data than the original. The key to this apparent magic is information theory.

Why Does It Matter?

In the decade from 2000 to 2010, the technology giant, Cisco, estimates that internet traffic grew by a factor of more than 300. Bear in mind that 2000 wasn't the prehistory of the internet; in fact it was the height of the dot-com boom, when many of us were already online both at home and at work, downloading music, shopping, sending email and getting into fruitless arguments on forums. Today, we expect to be able to stream movies on our home broadband, while companies move vast quantities of data between continents in seconds.

Anyone who has worked with raw sound or video or with hi-res photographs knows these need a lot of storage space, so many common file formats on the internet use compression. Some, like MP3, are "lossy," meaning the compression causes

Not all forms of compression are lossless: the JPEG format makes files smaller by discarding information, which leads to loss of details.

information to be thrown away. Since this can't be reconstructed at the other end, the quality of the data is permanently reduced. Others, like LZW, are "lossless" and manage to reduce the size of the data by representing it in a more compact way.

Information theory allows us to understand and quantify what's preserved by these algorithms, and what's lost in others. It was developed by Claude E. Shannon in the 1940s, although some progress had already been made towards it in the decades before that. Like a lot of early work in communications technology, it was done at Bell Laboratories in the United States at a time when the company held a virtual monopoly on the telephone business and invested heavily in research.

Beyond the internet this theory has been used productively in neuroscience, genetics and many other fields. As early as 1956 Claude Shannon was moved to publish a paper, "The Bandwagon," in which he urged researchers to be more cautious about trying to apply it to every problem that

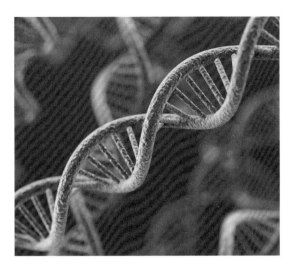

Your DNA fits something seemingly very complicated — the instructions for making you — into a surprisingly small amount of space.

crossed their desks. Nevertheless, his ideas have been applied very widely.

In More Detail

Shannon's key insight begins from the observation that determining the amount of information in a message isn't always as simple as counting the number of letters, numbers or other symbols it contains. An obvious example is a message that's repeated, as in some announcements on public transport systems — we certainly don't get twice as much information just by saying the same thing twice! This kind of redundancy can be very useful in a message precisely because it doesn't add new information but allows us to check the consistency of the information we received (see Error-Correcting Codes, page 130).

What's more, it seems as if a "complicated" string of numbers like AABAAAABABBBABBA contains more information than one with a repeating pattern like ABABABABABABABAB. The second one has some redundancy that could be simplified, while the first one doesn't. That's just the sort of thing the LZW algorithm exploits (it might compress the second string to "AB × 8"). In a sense the first message contains more

information than the other because it's less predictable, making it harder to describe in terms of repeating patterns.

This suggests that unpredictability is a good measure of information. In the extreme case, it seems that if I already know what you're going to tell me, your message communicates no information at all. If your message is highly unpredictable, then I have to pay close attention to it, because each new part is really new and unexpected. This suggests that entropy, the level of disorganization in the message, might be a good measure of its information content (see Entropy, page 74). Low-entropy messages can be compressed more than high-entropy ones because their information content is lower.

Suppose our messages consist of streams of symbols taken from a fixed "alphabet," which could be letters, numbers, types of fruit or whatever as long as there's a finite number of them and we can tell them apart. For example, let's suppose a naturalist has been using this alphabet to report sightings of different types of animal:

$$x_1 \qquad x_2 \qquad x_3 \qquad x_4 \qquad x_5$$

$$(P) \qquad (B) \qquad (O) \qquad (S) \qquad (A)$$

Penguin   Polar   Orca   Seal   Albatross
           Bear

Over a period of time perhaps we observe that the following probabilities attach to reports of each of these creatures:

$$p(x_1) \quad p(x_2) \quad p(x_3) \quad p(x_4) \quad p(x_5)$$

$$0.4 \quad 0.05 \quad 0.15 \quad 0.25 \quad 0.15$$

Shannon's formula allows us to calculate the entropy of this information using the sum (see Zeno's Dichotomy, page 18) of the probability of each message times its base-2 logarithm (see Logarithms, page 36):

$$H = -\sum_{(i=1)}^{5} p(x_i)\log_2 p(x_i) \approx -2.07$$

What Have You Done For Me Lately: Technology

The value of H, and therefore the amount of information, becomes a larger negative number as the number of possible signals increases and as their probability distribution becomes more uniform, reducing our ability to guess which signal comes next (see The Uniform Distribution, page 162). According to the Shannon-Hartley Theorem, however, this levels out at a limit: for a particular type of channel characterized by its bandwidth and how much noise it's affected by, there's a theoretical maximum amount of information you can squeeze down it. That means that at some point clever compression will stop working and we'll need to invest in higher-capacity channels with less noise if we want to move information around more quickly.

Shannon's definition of information is a technical one: it's precise but it doesn't necessarily agree with our everyday use of the term. For example, a random signal has very high entropy: a detuned radio receiving only static ("white noise") is, according to this definition, communicating much more information than one tuned to a radio station, even though the latter would seem to be much more meaningful than the former. Part of how we make meaning is by repetition of recognizable sounds, images and so on; part of what it means to be information, in Shannon's sense, is the avoidance of repetitive patterns.

This kind of thing often happens with scientific definitions, especially those definitions that claim to encapsulate an idea from everyday experience in a mathematical form. Almost always, the idea is transformed in the process and we have to be very careful not to switch between this new, precise version and the everyday version we're all familiar with.

Some information is best delivered in a compressed form.

Quantifying something as vague-sounding as "information" was an intellectual achievement in itself, and it has had profound effects on computing and communications technology.

Information Theory

The Fourier Transform

An alternative way of looking at functions that makes all our digital media (and many other things) possible.

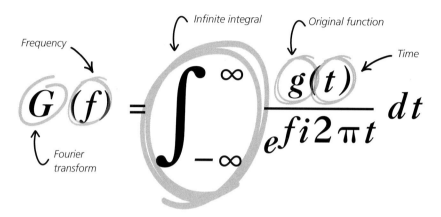

Frequency · *Infinite integral* · *Original function* · *Time*

$$G(f) = \int_{-\infty}^{\infty} \frac{g(t)}{e^{fi2\pi t}} \, dt$$

Fourier transform

What's It About?

The Fourier Transform is like a currency exchange for mathematical functions (see Logarithms, page 36). Suppose you have some function $g(t)$, which is awkward to deal with; in the analogy, just imagine $g(t)$ is a large-denomination note that the local stores are refusing to accept. Well, if you take $g(t)$ up to the cashier's window, they'll happily change it for you into a big bag of coins, each of which is easy to spend wherever you like. If you don't end up spending the money, you can return to the bank and they'll change your coins back into the note you started with.

In more grown-up terms, the Fourier Transform can convert almost any complicated function into an infinite collection of extremely simple ones. These simple functions are always the same — they're variations on the sine and cosine (see Trigonometry, page 14) — so the same tools can be used to manipulate them regardless of how weird the starting function $g(t)$ happens to be. Another analogy would be this: the Fourier Transform translates quirky, awkward functions into a single, universal language. We can then have a conversation with them in the universal language and, if necessary, translate the results back again.

Why Does It Matter?

Many technological devices have to deal with input from the outside world and, as we all know, the outside world can be a messy place. Very often the input comes in the form of a function, $g(t)$, and we can't really predict ahead of time what that function will be like. It would be very complicated to build systems that could deal with every possible function, and probably those systems would be fragile because nobody could guess exactly what "every possible function" ought to mean. For example, a CD player would need to contain masses and masses of software to deal with all the different waveforms that might appear in music and it probably still couldn't cope with some of those really weird records you sometimes put on.

The Fourier Transform means we don't have to do this. Instead, whatever signal we want to deal with is transformed into an infinite collection of simple trigonometric functions. The software does its thing with those functions and, if necessary, translates the results back again. Simple, sleek and effective. What's more, the Fast Fourier Transform (FFT), an algorithm invented (effectively) in the 1960s, makes it possible to make these changes in very quick time.

The Fourier Exchange swaps functions of time for functions of frequency and vice versa.

The field of Fourier analysis began long before this, with attempts to solve the equations governing vibrating strings (see The Wave Equation, page 84). That idea involved adding trigonometric functions together to produce different kinds of vibration. In a way it's miraculous that this idea has such far-reaching applications. After all, it may not be surprising that a violin string should vibrate in ways that are combinations of oscillating up-and-down motions, but it's not at all obvious that a wide range of non-repeating, sometimes jagged series of data should succumb to the same method. Yet they do, and that's the Fourier Transform.

In More Detail

When the cashier at the Fourier Exchange handed you that bag of simple functions in exchange for your complicated one, I mentioned it was a big bag and I meant it: in fact it's infinitely large. To pick one of these simple functions, we choose the frequency we want to use, f, and we get back $G(t)$, the Fourier Transform for that frequency. The frequency can be any number at all; which numbers we pick, and how many of them we need to sample to get an accurate enough picture of things, depends entirely on the application.

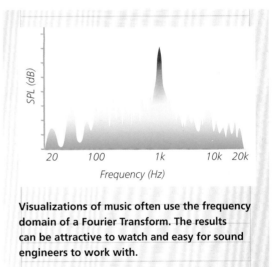

Visualizations of music often use the frequency domain of a Fourier Transform. The results can be attractive to watch and easy for sound engineers to work with.

It's now time to take the equation apart and see what makes it tick. What we have on the right is mostly a fraction, with our original function $g(t)$ being divided by something that looks pretty messy. Let's focus on the bottom of that fraction. At its heart is the fact (see Euler's Identity, page 40) that

$$e^{it} = \cos t + i \sin t$$

so this packs together the two trigonometric functions we're interested in. The number t is allowed to range over all the ordinary positive and negative numbers, which is what those rather alarming-looking infinity signs mean in the integral.

In our equation we have a bit more, though. Let's add the next part:

$$e^{i2\pi t}$$

Adding that 2π means that this whole thing is equal to exactly 1 when $t = 0$, and it repeats its oscillating cycle back to 1 over and over again whenever t hits a whole number. This is a way of getting things under control and making the function speak our universal language. Here comes the next part:

$$e^{fi2\pi t}$$

Here we're bringing in the frequency, which we choose when we pull one of these functions out of the bag. This is the only place we're bringing in f,

so it's the thing that makes a difference between our different functions.

Altogether, then, this is a periodic, oscillating function involving sines and cosines that we divide into our original function. We then sum up that whole thing across the entire range of possible values of t (see The Fundamental Theorem of Calculus, page 26). This should give us a number, which is the value of the Fourier Transform for a particular frequency f.

"Wait," you may be thinking. "Why should we expect to get a finite number when we add up all the infinite possible values of this thing?" This is not a foolish question. After all, I've been stressing that $e^{fi2\pi t}$ is a periodic function, one that repeats itself on a regular basis. What if $g(t)$ shoots off to infinity? Won't the integral be infinite then as well? The answer is, in part, that we're multiplying by a combination of periodic functions (sine and

The Fourier Transform makes Magnetic Resonance Imaging (MRI) work.

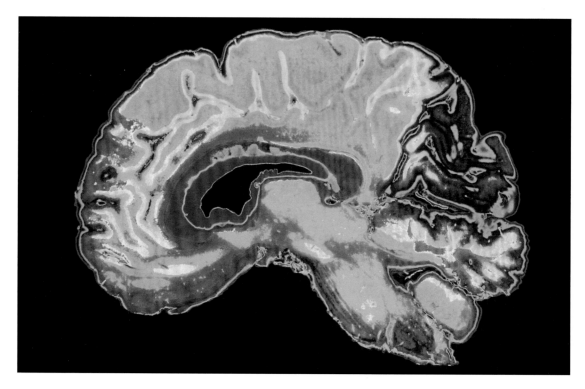

What Have You Done For Me Lately: Technology

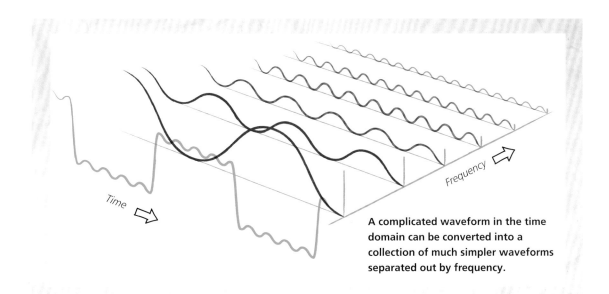

A complicated waveform in the time domain can be converted into a collection of much simpler waveforms separated out by frequency.

cosine) that are negative exactly as often as they're positive, which means the positive and negative bits really ought to "net off" to zero; the only elements that prevent this are the variations caused by the original function, $g(t)$. Those variations are just what the Fourier Transform captures.

You could see the Fourier Transform not as a new set of functions but as the original function looked at from a different perspective. The original function was in terms of time, which seems natural: usually a signal or set of data comes to us over time, and we store and reconstruct it in the same way. We like to see animations of mathematical processes evolving in time because that's the way we're used to navigating the world: allowing time to be the variable that changes and seeing what happens to everything else as it does.

Now, I said that the Fourier Transform gives us back a "bag" of functions that we access by frequency rather than time. The image of a bag suggests a disordered collection, but actually the frequencies are ordinary numbers arranged in a very orderly fashion, just like the instants of time we use to evaluate our original function $g(t)$. So in a sense all we've done is swap "time" for "frequency" as the window through which we're looking at the mathematical object we started with. Fourier Transform enthusiasts say it allows us to switch between the "time domain" and the "frequency domain." The time domain is more intuitively obvious to us, but in the frequency domain the math can be much simpler.

It was born of necessity during the development of the physics of heat and waves, but the Fourier transform now belongs close to the heart of calculus.

The Black-Scholes Equation

An equation that enabled the calculation of theoretical prices for options made digital, derivatives-based finance possible.

$$\frac{1}{2}S^2\,\sigma^2\gamma - \theta = r(V - \delta S)$$

- *Underlier's volatility* → σ^2
- *Theta* → θ
- *Gamma* → γ
- *Risk-free interest rate* → r
- *Delta* → δ
- *Option value* → V
- *Underlier's price* → S

What's It About?

In the financial world an "option" is a contract that allows you to buy or sell some quantity of a particular thing at a specified price on a particular day in the future. Options are transferable, meaning if you have one and no longer want it you can sell it to someone else who gets the same rights you had.

Options usually apply to financial assets, but they're easier to understand in relation to something concrete. Suppose you run a second-hand car dealership; you buy a car from someone at a given price and aim to sell it at a higher price. You lose money if the price people are willing to pay for that car drops, and in fact you stand to lose all your money if nobody's willing to buy it at any price. This could happen, for example, if that particular model was revealed to have a serious safety flaw.

You don't mind a little risk, but the possibility of losing everything because of a price change that isn't even your fault seems a bit extreme. So you buy an option to sell the car at a price below what you paid for it, but not so much below that you'd take a huge loss. The option is cheap because nobody wants to sell that type of car that cheaply right now, so the person who created it doesn't think it's likely to get used — but if it is, they'll get the car at a knockdown price, so they're happy. If car prices change dramatically, though, it might just save your bacon and force the person who sold you the option to buy it at a price that doesn't look so attractive any more.

Like most derivatives, then, an option is a sort of insurance contract. You buy it and you might never use it, but the cost is offset by the fact that it might save you a lot of money. As you probably know, because options can be sold on, they're also traded speculatively, a practice that's often rather unfairly blamed for all the ills of the financial world.

Why Does It Matter?

The trouble with an option is working out what it's worth. That depends on the details of the contract, in particular the price it allows you to sell the car for and when it expires. It also depends on things outside the contract, especially the going price for a car of that type. It seems as if there ought to be a formula that relates all those things to each other and gives you a fair price for the option, but it's not obvious how to put it together.

A big part of the problem is that we're looking

into the future. What will the car's market value be in a year's time? We don't know. That's partly why someone might want the option in the first place, but it makes coming up with a fair price rather tricky.

In the early 1970s American economists Fischer Black, Myron Scholes and Robert Merton used some fairly exotic mathematical ideas to solve the problem. Their equation rests on a lot of assumptions, but it represented impressive progress. The Black-Scholes Equation also gave birth to modern mathematical finance and became the basis for understanding almost all derivatives products. It's not much of an exaggeration to say that, for better or worse, our contemporary financial world, driven by high-powered computers and even-higher-powered mathematics, owes its existence to this pioneering work.

In More Detail

To make sense of the equation we'll need to decode that forest of Greek letters, each of which stands for an important ingredient of the option-price recipe.

Let's work with an option that gives you the right to sell a particular thing — we call that thing the "underlier" — at a fixed price, the "strike price," on a fixed date in the future (this last condition makes it a "European-style option"). You don't have to use it, just like having travel insurance for a trip but not claiming on it.

Suppose we have a certain amount of the underlier and we buy an option to sell it at a fixed price. If the market price of the underlier goes up, we'd expect the market value of the option to go down, because it becomes less attractive.

It seems, then, that we can arrange things so that we buy just enough options for their price fluctuations to exactly balance out the underlier's

The Chicago Board Options Exchange opened in 1973 and quickly became one of the first testing grounds of the Black-Scholes model.

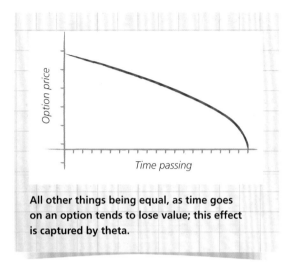

All other things being equal, as time goes on an option tends to lose value; this effect is captured by theta.

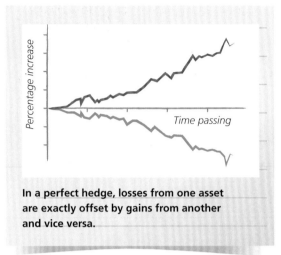

In a perfect hedge, losses from one asset are exactly offset by gains from another and vice versa.

price changes: if we make money on the underlier we'll lose exactly the same amount on the option, and vice versa. All other things being equal, this is a "perfect hedge": we get to hold onto the underlier without the price changes affecting us at all.

How many option contracts do we have to buy? That's determined by how the option price varies with the underlier's price, a quantity known as "delta" (δ):

$$\delta = \frac{dV}{dS}$$

If you own one unit of the underlier, so its value is S, you need options to the value of δS to hedge the position. Overall, then, we might expect the theoretical value of the option to be the same as the value of the hedge, that is

$$V - \delta S = 0$$

but this isn't always the case, because changes in the underlier's price aren't the only factors to consider. That discrepancy is what the left-hand side explains. Incidentally, in the equation the right-hand side is multiplied by a "risk-free interest rate," r — that's important for getting the price right in real-world applications but for our purposes we can safely ignore it.

On the left-hand side two elements are added together; their total should equal the difference between the option value needed to hedge out price changes in the underlier and the actual option price we expect to see in the market. This difference comes from two sources.

The easier to understand is theta (θ), which measures the tendency for the passage of time to affect the option's value:

$$\theta = -\frac{dV}{dt}$$

After all, time is built into the option contract: the option comes into effect on a specific day. As that day gets nearer, the chances of the underlier's price changing enough to swing the value of the option decreases. This is one factor that explains why we might pay a different price for the option than simply δS, the price implied by the need to hedge out price movements in the underlier.

The other term is more complicated and involves gamma (γ), a measure of the "acceleration" of the option price when the underlier's price changes, which is the same as the rate of change of delta:

$$\gamma = \frac{d^2 V}{dS^2} = \frac{d\delta}{dS}$$

What Have You Done For Me Lately: Technology

As we get closer to the expiry date, a change in the underlier's price will have a proportionally greater effect on the option's price, since the element of uncertainty is reducing; this is a sort of counterbalance to the effect of theta. We multiply this by a factor describing how much the stock price *does* actually tend to change.

Putting all that together means adding the theta element, which captures the direct effect of time on the option price, to the gamma element, which captures the effect of time on delta. When we do that we get the left-hand side, which (as you might recall from something you read earlier) represents the difference between the option's price and the price implied by the "perfect hedge." Solving this equation for V — not a straightforward task! — gives the theoretical "fair price" for the option.

Mathematical models of real-world events always involve assumptions, simplifications and approximations — sometimes rather drastic ones. The more complicated the model, the harder it can be to see exactly what those are and when they're likely to cause trouble. In the end, any equation is just an equation; its effects depend on the judgments of the people and institutions that use it.

Using your insurance policy might not be ideal but it beats not having one.

Modern finance wouldn't be the same without the Black-Scholes equation, which brought high-powered mathematics and physics into the world of asset pricing.

The Black-Scholes Equation

Fuzzy Logic

From ancient philosophy to air-conditioning systems, sometimes the solution to a problem is to be a bit less precise.

Not

$$\neg x = 1 - x$$

And   Lesser value

$$x \wedge y = \min(x, y)$$

$$x \vee y = \max(x, y)$$

Or   Greater value

What's It About?

Usually we like to think our beliefs are made up of statements about the world that are either true or false. Either my neighbor has a dog or she doesn't; it can't be both and it can't be neither. There are really only three things we can say: the dog exists, the dog doesn't exist or I don't know. If it's the

last one, or I have a hunch but I'm not certain, we enter the realm of probability (see The Uniform Distribution, page 162). The fact is, though, that

Rain isn't something that's simply on or off: it comes in degrees and it often makes sense to say it's raining a lot, a bit or hardly at all.

Wet   Dry

146

What Have You Done For Me Lately: Technology

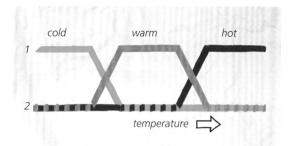

Fuzzy logic makes environments with controlled temperatures more pleasant to be in.

either the first or second of those possibilities is actually true and the other is false.

This is classical logic: a logic that admits two truth-values and insists that every statement must have one and only one of them (see De Morgan's Laws, page 126). It's a black-and-white, crystal-clear way of thinking that doesn't always fit real-life situations terribly well.

For example, suppose I say it's quite warm today. If it's very cold, you may say my statement is false; likewise, if we happen to be in the middle of a heat wave. Between these two extremes, though, it's not entirely clear what happens. We may feel that classical logic forces us to say that, for every precise temperature it could be, the statement "It's quite warm today" is either true or false. This is a bit ridiculous; surely there are some temperatures where it's somewhat true, or slightly true, or very true?

Fuzzy logic, invented in the 1950s, offers a way to capture this mathematically. Where classical logic can be represented by two numerical values, 0 for false and 1 for true, fuzzy logic allows us to use all the values in between them, just as we do in probability. It allows for degrees of truth and falsehood.

Why Does It Matter?

Some things in life really are cut and dried, and classical logic does a great job of dealing with them. An example is whether an ordinary light (not on a dimmer!) is on or off — either it is or it isn't, and any prevarication about that is probably best left in the philosophy seminar room. Other situations aren't so crisp, but we can simplify them in a way that makes them so without doing great violence to them. An example would be having a job. There are some borderline states when you might not be sure whether "I have a job" is a true or false statement. You might be about to be made redundant, working out notice, doing a short gig as a contractor and so on. Yet we can probably tidy those things up with a few official definitions without going crazy and throwing standard logic out of the window.

In other cases, though, that approach doesn't work so well. The most famous one is a heating and air-conditioning system controlled by a thermostat. In the naive design, when the temperature drops below a certain level, it's true that it's cold, so the heating system turns on; when it goes above that level it's now false that it's cold: the heating turns off and the air-conditioning starts blowing cold air to cool us down. Result: the temperature swings violently between hot and cold, which can be uncomfortable for everyone. Replacing this with a system using fuzzy logic creates a much smoother experience, and in fact that's how most modern temperature-control systems work.

This general principle can be found in many technologies that need an automated system to adjust the way they work, from mass-transit systems, aircraft auto-pilots and satellites to home electronic goods of all kinds. At a more conceptual level, it can be helpful for scientists working with intrinsically indistinct phenomena and it's even been embraced by some philosophers. It can also help improve decision-making in law and medicine, especially where "expert systems" are automating parts of those processes.

In More Detail

Fuzzy logic attempts to answer a problem that dates back to the ancient Greek philosophers. There are many forms: today it's most often called the problem of vagueness. Here's a modern example: suppose we both agree that, for a particular stretch of road, driving at 15 km/h

(10 mph) is not dangerous and driving at 160 km/h (100 mph) is dangerous. Suppose we also agree that a dangerous speed is always faster than a not-dangerous one: that is, you can never drive more safely by speeding up, all other things being equal.

Well, then, there must be a cut-off speed somewhere between "not dangerous" and "dangerous," mustn't there? To see why — and why this is a problem — imagine driving down the road starting at 15 km/h (10 mph) and accelerating smoothly up to 160 km/h (100 mph). At the beginning our speed was not dangerous, and it stayed not-dangerous for a while. At the end, it was dangerous. So, at some point, it must have switched over. That point, if we could find it, would be a really good guide to what the speed limit should be around here.

Still, can it really be true that at some point

One moment, the airplane is in the cloud, and then it's out of the cloud. But is there a precise moment when the transition from "inside" to "outside" occurs?

during this experiment we were driving at a perfectly safe speed, and just a fraction of a second later we'd accelerated (barely noticeably) to a dangerous one? That seems silly. In fact, if the speed we were going at was safe, going a tiny fraction faster should be safe too, shouldn't it? And if we're driving at a dangerous speed, it can't be that slowing down by a microscopic amount can make it safe.

The trouble here is that "dangerous" is a vague word. If we try to give it a precise definition, such as "driving at more than 50 km/h (30 mph)," we lose the original meaning and turn it into a technical term. The law does this all the time and, although it often works well enough, it can create grave cases of injustice when someone falls just the wrong side of a fairly arbitrary line that doesn't seem to map onto the concept it's supposed to be capturing.

This is where fuzzy thinking can step in. Before, we had a statement $A =$ "This speed is dangerous," and we could check its truth value: $T(A) = 1$ if the speed is dangerous and $T(A) = 0$ if it isn't. Now, we allow $T(A)$ to take values in between. In the case of

What Have You Done For Me Lately: Technology

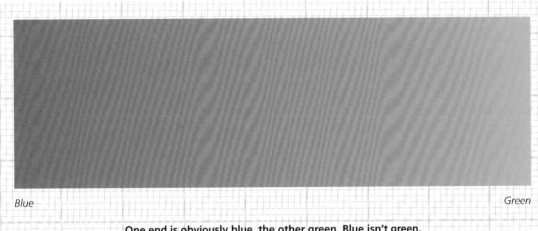

Blue　　　　　　　　　　　　　　　　　　　　Green

One end is obviously blue, the other green. Blue isn't green, so it seems there must be a precise point where the color changes from blue to not-blue. Where is that point?

the speeding car it would seem natural to have $T(A)$ = 0 at 15 km/h (10 mph), $T(A)$ = 1 at 160 km/h (100 mph) and in between let it increase in some smooth way. In that case, we wouldn't say, "This speed is dangerous" but something more like, "This speed is 0.8 dangerous." The example exposes a downside of fuzzy logic: it's hard to imagine road signs giving information like this, rather than just a flat speed. But we can perhaps imagine penalties being applied in this way, and to an extent they already are, for doing 160 km/h (100 mph) in a quiet neighborhood will likely get you into a lot more trouble than going just a little over the limit.

If this is going to be a logic, it needs more than just truth and falsehood: it needs logical operators like "and," "or" and "not" (see De Morgan's Laws, page 126). The most frequently used options are the so-called Zadeh operators given above: instead of "A and B" being true when A is true and B is true, we say "A and B" takes the minimum of the (now numerical) truth values of A and B. A chain is, after all, only as strong as its weakest link. These operators allow computers to apply fuzzy logic in complicated situations involving many variables, and with the same kind of power they bring to the black-and-white problems of classical logic.

It's not easy to give up the reassuring binary world in which everything is either true or false, but when we do, we sometimes find problems much easier to solve.

Degrees of Freedom

This concept is central to robotics and provides a neat application of high-dimensional spaces.

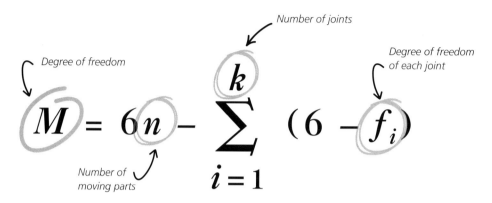

Number of joints

Degree of freedom

k

Degree of freedom of each joint

$$M = 6n - \sum_{i=1}^{k} (6 - f_i)$$

Number of moving parts

What's It About?

A helicopter hangs in the air. How can we describe the different ways it can move? Well, it can go forward and backward; we can use a single number to represent a motion like this, with positive numbers meaning forward and negative meaning backward, and the number itself being, say, the number of meters or yards it moves. Similarly, it could move from side to side, and for

this we need another number, say positive for right and negative for left. We can't combine these into a single number, since they're independent of each other. Finally, it'd be a sorry excuse for a helicopter if it couldn't go up or down, so we'll need another number for that.

We're not quite finished, though, because a helicopter can do more than that. For a start it can rotate to point its nose up or down — pilots call

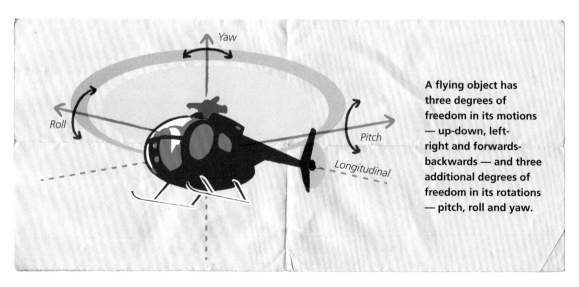

Yaw

Roll

Pitch

Longitudinal

A flying object has three degrees of freedom in its motions — up-down, left-right and forwards-backwards — and three additional degrees of freedom in its rotations — pitch, roll and yaw.

What Have You Done For Me Lately: Technology

this "pitching." Again, this is independent of the direction it's moving (let's assume), so we'll use a fourth number for this, which can just be the angle of the pitch. As you might know, there are two other kinds of rotation it can do: "roll" and "yaw," and we'll need separate numbers for those too (I'm not sure rolling is a good idea in a helicopter, but what do I know?). Altogether, then, a helicopter can move in six independent ways, so we say it has six degrees of freedom.

In More Detail

The helicopter's position, or a particular movement it makes, can both be specified by a list of six independent numbers. This is a bit like specifying a point in space using its coordinates, only now we seem to have rather a lot of them. This doesn't alarm mathematicians, though, because these points in space can be manipulated very easily using something called linear algebra, and linear algebra doesn't care very much how many dimensions there are. So, next time you see a helicopter, point it out to your friends and say, "Look, a six-dimensional space!"

For an object in three-dimensional space, that's as good as it gets: three directions of movement and three of rotation. This is, for example, the standard way to specify an object's position or movement in a computer visualization or game, though the rotations can be troublesome (see Quaternion Rotation, page 152). Things get more interesting when we have systems that are constrained in various ways, like mechanical arrangements of beams connected by joints to make a robot arm.

The joints may restrict the degrees of freedom in various ways. As a mundane example, consider

Each joint in the robot arm increases the total number of ways it can move; joints of different types add different numbers of freedoms.

a door: as a wooden rectangle it has all the freedom of a helicopter to move around as it likes in space, but once attached to a doorway by hinges it suddenly has only one degree of freedom, its "yaw." It can't move by changing its position at all, and while it can still rotate it can only do it in one way. The hinge, as a type of joint, allows only one degree of freedom. It has single-handedly collapsed the six-dimensional space into a one-dimensional one: a circle representing the possible angles the door can be at.

A flying plane is, in a sense, a six-dimensional space. Surprisingly, thinking of it in this way is sometimes actually useful.

Degrees of Freedom

Quaternion Rotation

A mathematical curio of the 19th century solves a range of practical problems in the 20th and 21st.

$$\overset{\curvearrowright \sqrt{(-1)}}{\boxed{i^2}} = \overset{\curvearrowright \sqrt{(-1)}}{\boxed{j^2}} = \overset{\curvearrowright \sqrt{(-1)}}{\boxed{k^2}} = ijk = -1$$

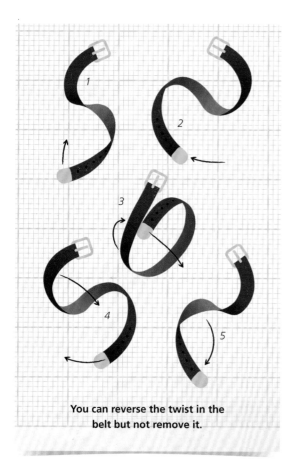

You can reverse the twist in the belt but not remove it.

What's It About?

Try the following experiment with an ordinary leather belt and a friend. You each hold one end of the belt, with the end horizontal to the ground and pointing straight at you. Put a single twist in the belt by turning one end through a complete circle. The challenge is to undo the twist by moving the ends of the belt but always keeping them parallel to the ground and pointing in the same direction (no rotations whatsoever).

The two of you can move your ends of the belt around as much as you like, and what you'll find is that you can reverse the twist in the belt, but never untwist it. That's odd: reversing the twist seems to involve a *double* twist in the opposite direction to the original one; but if so, you must at some point have completely removed the original twist. Try as you might, though, you'll never find the place where that happens, because it doesn't.

This phenomenon suggests that turning things around in 3D space is more complicated than it seems and might not always work the way you expect. The quaternions are exotic mathematical objects that seem to capture how these rotations *really* behave, although they don't look anything like rotations when you first meet them.

What Have You Done For Me Lately: Technology

Why Does It Matter?

We live in three-dimensional space and many physical problems involve rotations in that space. How can we describe them? First, set up a coordinate system representing the three directions — forward-backward, left-right and up-down. Now imagine rotating around each of these directions (see Degrees of Freedom, page 150). Rotating around the forward-backward direction would send you head over heels in a sideways, cartwheeling motion. Rotating about the left-right direction would also send you head over heels, but this time in a forward somersault. Rotating about the up-down direction is like standing on the middle of a roundabout. Rotations like this are sometimes called "Euler angles" and combining them can produce any other sort of rotation you can imagine.

This works well, at least until it doesn't. Euler angles didn't work so well on May 18, 1969, when the Apollo 10 lunar module was trying to dock with its command module in preparation for a return to Earth. When trying to execute the rotations required for the docking maneuver, the astronauts lost control of the module because of a phenomenon called "gimbal lock." In the guidance system, two of the three rotations were effectively synchronized or "lined up" so that they represented the same rotation. The astronauts were forced to carry out a large, unplanned manual maneuver to shake the system out of its deadlock and regain control. The same very dangerous situation can theoretically affect any kind of aircraft.

A gimbal is a mechanical device, but this phenomenon is also well-known to video-game developers. Although Euler angles are intuitive and easy to work with, the extreme rotations in many games can easily create gimbal lock and similar unpleasant situations. Today, then, spacecraft, aircraft and even video games are all

Apollo 10 returned home safely after an alarming brush with the phenomenon of gimbal lock.

designed with quaternions instead of Euler angles. Quaternions are also useful wherever symmetries of three-dimensional objects are of interest, so some molecular biologists and chemists use them and they're even found in the Pauli matrices that define quantum spin. What's more, in a sense you use them yourself almost every time you move around in three-dimensional space (which you should try to do every day if you can).

In More Detail

The quaternions are usually introduced as a special number system. They're rather like the complex numbers (see Euler's Identity, page 40), which come from adding a single special number, i, to the ordinary numbers to act as the square root of −1. The idea of the quaternions is to add two more special numbers to this system, different from each other but also square roots of −1. If this seems outrageous, remember that even in our ordinary number system the number 4 has two square roots, 2 and −2, since $2^2 = (-2)^2 = 4$. So multiple square roots aren't weird in themselves. These new numbers are called j and k. The extra rule, given

in the equation, is that when we multiply them all together we also get −1.

Every number in the quaternion system is some ordinary-number multiple of each of four basic elements — 1, i, *j* and k — added together. We can represent this as a list or, as mathematicians say, a "vector." For example, here's a quaternion:

$$(3,-2,1.8,-3.72)=3-2i+1.8j-3.72k$$

If you bring together all the possible lists we can make like this, that's the quaternions. Notice that there are four independent entries in the vector, so a quaternion is a four-dimensional object.

Does this actually work as a number system? After all, we can't just invent new numbers: we have to be able to do sensible things with them like adding, multiplying and so on. It turns out that we can define these things, although the rule we get for multiplying is particularly messy. Still, the quaternions make quite a "nice" number system, as these things go: it's an example of something called a "real finite-dimensional division algebra." The only possible examples of this kind of structure

A gimbal controls three rotations in space, but can get "locked" so that one of them becomes unavailable.

What Have You Done For Me Lately: Technology

are our everyday numbers, the complex numbers, the quaternions and an extension of them called the octonions.

To relate quaternions to rotations we begin by restricting ourselves to unit quaternions, which means those whose "size" (measured using Pythagoras's Theorem [see page 10]) is equal to 1. This turns our four-dimensional quaternions into three-dimensional objects, although that's not at all obvious in vector notation. Still, it's reassuring: three-dimensional rotations ought, I'd say, to be three-dimensional things. And it turns out that, after some slightly scruffy-looking algebra, we can convert unit quaternions into rotations of three-dimensional space.

Even if you believe me about the algebra, you might wonder what's been achieved: haven't we just found a different way to represent the same objects? It's true, we have: but representations are important because we do things with them. There's something fundamentally wrong with Euler angles as a representation of rotations, but that wrongness only comes out when you start using them to do things. Do one thing and you won't see problems, but start doing one after another — as mathematicians say, "composing" rotations — and things can go awry. This doesn't happen with quaternions because that fundamental flaw has been fixed.

So what is this fundamental flaw, exactly? Well, it involves a rather deep and beautiful interplay of geometry, topology and algebra. Here's a glimpse of it: if you represent rotations with Euler angles they become, topologically speaking, a torus or donut shape (see The Euler Characteristic, page 44),

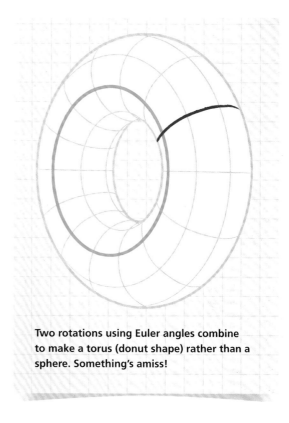

Two rotations using Euler angles combine to make a torus (donut shape) rather than a sphere. Something's amiss!

whereas if you use quaternions they become a sphere. Rotations are more sphere-y than torus-y, so in a sense, Euler angles act as a map that's the wrong shape. Think of the way, with a flat map of the Earth, you always have to make a sudden jump when you get to an edge; that's because the Earth is sphere-y, not flat-plane-y, and it's a similar sort of problem (see The Mercator Projection, page 110). To say more would get us into Lie groups and covering spaces: topics for another day.

Rotations aren't quite what we usually think they are — most of the time that's not a problem, but just occasionally, it catches us out.

Quaternion Rotation

Google PageRank

The founders of Google made their fortune by solving a very big equation.

Rank of page A

Number of links to A

Rank of linking page

$$R(A) = 1 - d + d \sum_{i=1}^{n} \frac{R(T_i)}{C(T_i)}$$

Damping factor

Links on linking page

What's It About?

Google describes itself as providing an index to the World Wide Web. Yet there's a problem: there are a few thousand commonly used words in any given language, but there are billions of web pages. Any search engine that just gave you back all the pages that contain the word you're looking for, in jumbled-up order, would be a disaster, swamping you with choices, many of which are of poor quality. What's needed, among other things, is some way to work out whether a page is "good" or "bad." That can't be done by hiring someone to look at them all — there are too many — so we need some measure that a computer can apply on its own. Yet the computer can't *understand* what's on a web page, so how can it make that judgment?

The idea behind PageRank is that it doesn't have to understand, at least not directly, because other people will already have done it in a way the computer can interpret. When someone creates a page they often link to other pages belonging to other sites, and presumably they do this because they think those pages are interesting or valuable. A computer can be programed to sit there and collect information about these links, figuring out who links to whom in a vast, unimaginably complicated network of pages. A page with a lot of incoming links has already been "voted" more important than one on the same topic with only a few.

At least, that's the theory. The problem is that it matters who links to your page. If it's a well-established, credible site, then you should score more than if it's just another little unknown one. It might even be a spam site that exists solely for the purpose of linking to another site and thereby boosting its credibility (this was a real problem at one time). So how can the computer make a judgment about which of those linking sites are the

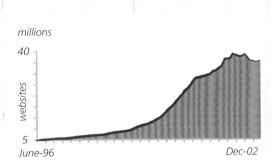

millions

40

websites

5

June-96 .... Dec-02

The number of websites went through a period of exponential growth and now poses a major problem for anyone seeking to make sense of it.

most credible? Easy — use the number of incoming links they have!

But now, perhaps, you see the problem: we seem to be stuck in a vicious circle. We need to know the credibility of the links coming into your site to evaluate it, but to find those out we need to know the credibility of all those linking sites; but to find those out, we need to repeat the process again. So it goes on, but since the internet is a

finite thing we must eventually end up back where we started. It seems like an impossible situation, but this is the problem the equation for PageRank solves.

Why Does It Matter?

Presumably, you know what Google is and why it's important to a lot of internet users. At the time of writing it has a near-monopoly on general-purpose internet searches. At the turn of the millennium most of us used several different search engines and we passed around arcane knowledge about which we thought were superior for which sorts of tasks. Google achieved dominance in that crowded market by providing higher-quality search results than anyone else at the time. This they did, in part, by means of PageRank.

In More Detail

First of all, we should say first of all that Google, like other search engines, uses a combination of

The amount of data handled by Google's servers has grown explosively. Finding our way around it is likely to be an ongoing problem.

A mini-internet with just three pages, along with the matrix used to calculate their PageRanks. The actual values depend on the damping factor chosen.

multiple different strategies and most of that is commercially sensitive information. We're only looking at a small part of that here: the calculation of a single number that feeds into the decision-making process. By the time you read this book, the way Google does things might have changed, but the mathematical principle will still be just the same.

The basic formula is very simple. We're trying to calculate PageRank for page A. On the right we have a big sum (see Zeno's Dichotomy, page 18) over all the pages on the internet that link to page A. For each one, we count how many times it links to page A. Then we divide by the total count of all links going out of that page. Why? Because a link from a page that's very choosy about what it refers us to is worth more than one from a page with thousands of indiscriminate links.

In fact the right-hand side is the *definition* of the PageRank of a page A; the problem is that PageRank appears there, too. That's like the thing your English teacher used to tell you off for: defining a word using the word itself. It's like saying

a person is "nice" if they have nice friends. It's that vicious circle that causes all the trouble. We need some sort of tool that can break the deadlock, and that tool is linear algebra.

Among other things, linear algebra deals with matrices. Think of a matrix as a square of numbers. You can find an individual number in matrix L by looking at the ith column and the jth row, just as you might remember the location of your car in a large parking lot, and by convention we write the number in that position as L_{ij}. Our matrix will be huge: it will have one row and one column for every page on the internet. If page T_i links to page T_j then the matrix entry L_{ij} will be 1/C, where C is the total number of links going out of page T_j. If the pages don't link, the entry will be 0.

This matrix helps us calculate PageRank in the following way. We'd like to have a list of all the PageRanks of the pages in our collection. If we turn this list into a vector called P, we will discover that it must satisfy the following innocuous-looking equation:

$$P = LP$$

If you know how to multiply a vector by a matrix you might like to grab a pencil and paper and see for yourself that this is exactly equivalent to the

What Have You Done For Me Lately: Technology

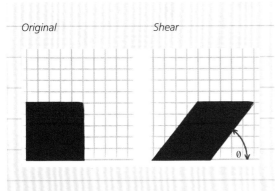

Original   *Shear*

This matrix transforms the square into the diamond shape, but the points along the bottom edge remain exactly where they were; these are the matrix's fixed points.

definition at the top of this section — the details are essentially simple but too grisly to lay out here.

Although I claim this is exactly the same as the definition we started with, we've achieved something nevertheless. The fact that P (our set of PageRanks) appears on both sides of the equation is no longer a problem. Multiplying P by L doesn't do anything; in fancy terms, this makes it a special example of an "eigenvector" of L.

If you've never seen this kind of algebra before it might look like voodoo. Really, though, it's just an example of a general sort of problem, and you've probably solved other examples in the past. For example, suppose we have the function

$$f(x) = 6 - 2x$$

Now, the equation $P = LP$ is like the equation

$$x = f(x)$$

Is there a vicious circle here? Not really. We can replace $f(x)$ with its definition and do a spot of symbol-shuffling to find a value of x that makes it true:

$$x = 6 - 2x$$
$$3x = 6$$
$$x = 2$$

The fact that our equation involves multiplying a vector by a matrix doesn't really make it any more complicated. Ultimately, it comes down to a lot of simple arithmetic. No, the problem is that L is a square containing about 625,000,000,000,000,000,000 numbers representing links, while P is a vector with about 25,000,000,000 PageRanks, so multiplying them together isn't something you want to try with a pocket calculator.

The good news is, there are some very clever methods for finding and working with the eigenvectors of matrices because these things have lots and lots of other applications. Long and boring those methods may be, but computers don't mind doing a lot of mindless calculations, and that's all that's involved in finding P.

We started off with a small problem of finding the PageRank for a single page, but here we solve at one swoop a much larger and more difficult-seeming problem: we compute the vector of all PageRanks at once. A lot of linear algebra is like this: a seemingly daunting series of tasks suddenly collapses into something easier, as if by magic.

For this reason, among others, matrix algebra is a basic tool in areas far beyond the world of search engines. In fact, it's hard to think of any part of technology where the eigenvalues of a matrix aren't an important thing to know about.

PageRank is one company's attempt to make sense of the internet, and it serves as a nice example of the power of linear algebra.

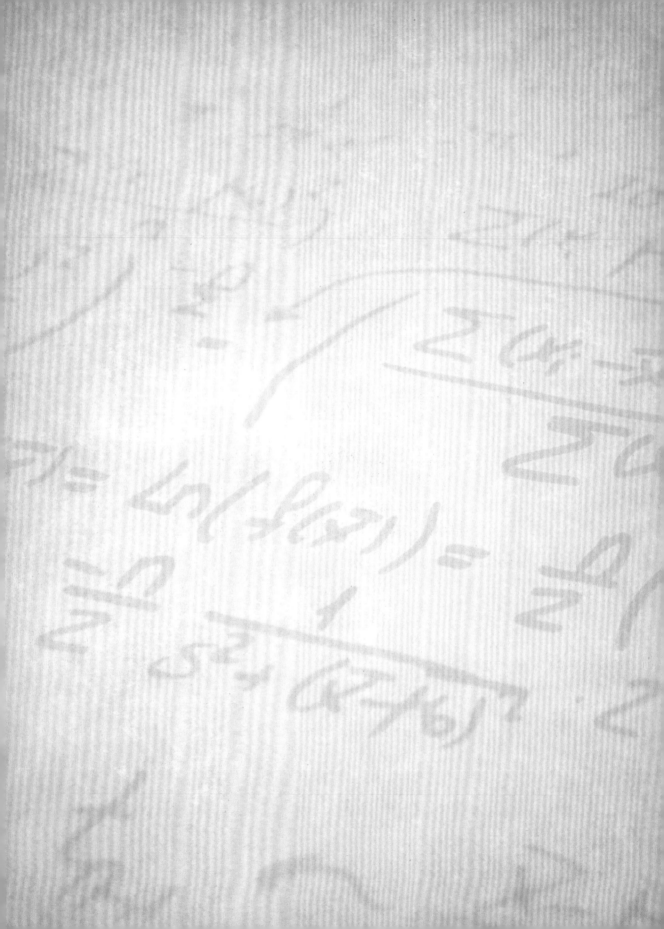

Known Unknowns

CHANCE AND UNCERTAINTY

The Uniform Distribution

The Uniform Distribution underlies games of chance and is the starting point for Bayesian Inference, which has connections with medicine, science and AI.

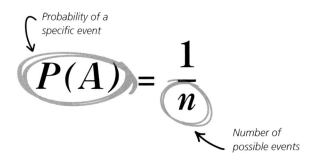

Probability of a
specific event

$$P(A) = \frac{1}{n}$$

Number of
possible events

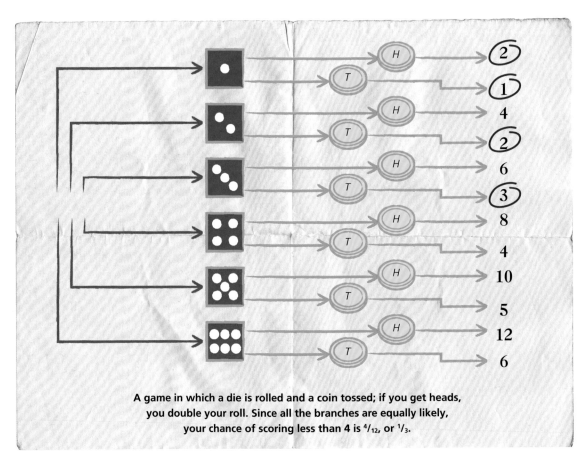

A game in which a die is rolled and a coin tossed; if you get heads,
you double your roll. Since all the branches are equally likely,
your chance of scoring less than 4 is $^4/_{12}$, or $^1/_3$.

Known Unknowns: Chance and Uncertainty

What's It About?

The theory of probability gives us a way to obtain knowledge about things that are uncertain. This is quite a feat, and for a long time it was widely believed to be impossible. The Uniform Distribution describes the simplest possible situation, in which we have a known, finite set of possible outcomes that are all equally likely.

Consider rolling an ordinary die. There are six possible outcomes and if the die is fair, and thrown in a fair way, there's no reason to expect one face rather than another. We say each face has an equal chance of coming up and that the whole set of (six) possible outcomes has the Uniform Distribution.

This isn't terribly helpful on its own, but with a little more work, we can attach numbers to those probabilities and use them to answer more complicated questions: how much more likely is it, for instance, that I'll roll an even number rather than a number less than 5? What if we roll several dice and add up the scores? From such humble beginnings, probability slowly grows into a precise theory of what we call "uncertainty."

Why Does It Matter?

Suppose you're in a math classroom and the teacher draws a triangle on the board. She points to one corner and declares that the angle there is a right angle, and asks you to solve some problem or other about it. Do you even think to ask how sure she is about that right angle? What if it's not exactly 90° after all?

Most likely you'll be told that it's assumed or defined to be a right angle for the purposes of the question. You can take it as given (another favorite mathematical term) and don't need to question it. Real life, though, isn't like that. In real life very little is certain right away, and even if we're sure of something we often have to admit there's a small chance we might be wrong.

Up until the 1600s, most mathematics had been based on certainty: geometrical problems, for example, in which lengths, angles or points were given and deductions were made on that basis. From information that's certain, it's possible to derive certain results. For example, given a right-

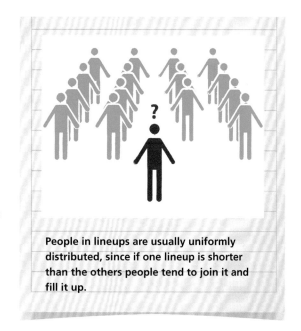

People in lineups are usually uniformly distributed, since if one lineup is shorter than the others people tend to join it and fill it up.

angled triangle and given that its two shortest sides are 3 units and 4 units respectively, we can conclude without a shadow of a doubt that its other side is 5 units (see Pythagoras's Theorem, page 10). But when, outside math class, has anyone ever given you information like this?

In many situations our information isn't absolutely precise, complete or certain. Scientific observations, for example, always involve a certain range of error; even taking that into account, it's always possible that the equipment is miscalibrated or faulty, or that mistakes have been made in calculations and so on. Almost everything we know is clouded by a little doubt, even the things we're pretty confident about.

Probability theory began as a way to understand games of chance played with cards and dice, where the Uniform Distribution reigns supreme. Since then it's become central to many fields of science and technology. Early in its history it was applied to physics and finance; later it helped give birth to more obviously statistical subjects like sociology and psychology. The sciences build mathematical models of the situations they're trying to understand; with probability these models often become far more powerful and effective.

In More Detail

In situations like flipping a coin, rolling a die or picking a card from a shuffled deck, we tend to assume that all the outcomes are equally likely — that is, that we're working with the Uniform Distribution. In the die-rolling scenario, we can ask what our chances are, in a given roll, of getting a 1; the answer is $\frac{1}{6}$ or, as we often say "one in six." This way of speaking is suggestive: what we mean is that if we roll the die 6 times, we expect it to come up 1 about once. It might not come up with a 1 at all, or it might even come up 1 every time (see page 162), but on average if we roll the die lots and lots of times we expect to get 1 on about one in six of those rolls.

Suppose that to win a game we need to roll either a 5 or a 6. This can be thought of as a new event: we must "roll 5 or 6." We can also see it as a combination of two simpler events: we must "roll 5" or "roll 6." It turns out that we can capture the behavior of the word "or" here by adding the two probabilities. Since either one will do for us, we combine them: $\frac{1}{6} + \frac{1}{6} = \frac{2}{6} = \frac{1}{3}$, so we have a one in three chance of winning.

If we play this game many times, for every three situations in which we end up needing to roll a 5 or a 6, we'll get it about once. This technique can be used to assign probabilities to any combination of outcomes joined by "or" as long as they can't both happen.

In the extreme, the chances of getting any one of the six outcomes is $\frac{1}{6} + \frac{1}{6} + \frac{1}{6} + \frac{1}{6} + \frac{1}{6} + \frac{1}{6} = 1$, and we describe an event with a probability of 1 as "certain." No probability can be higher than this. Assuming we disallow weird results like the die landing on one corner and balancing, it's certain that we'll get one of those six results. Similarly, we define an impossible event such as rolling a 7 on a normal die as having probability 0. Mathematically this works out: the chances of me getting either a 6 or a 7 is just $\frac{1}{6} + 0 = \frac{1}{6}$, the chances of me getting the 6. That's right because the extra option "or a 7" doesn't make any difference, since getting a 7 isn't going to happen.

When everything that can happen has an equal chance of happening, events are uniformly distributed.

Known Unknowns: Chance and Uncertainty

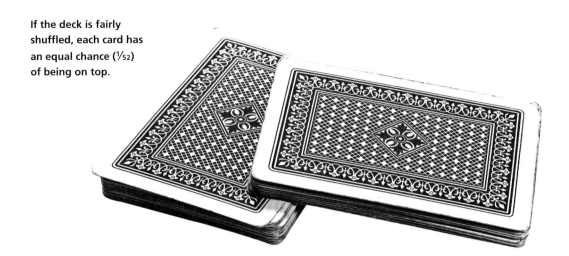

If the deck is fairly shuffled, each card has an equal chance (1/52) of being on top.

This gives us a trick that's often useful when calculating probabilities: the chances of some event *x* not happening are 1 − *p*(*x*), since the chances of it either happening or not are clearly 1! For example, the chances of rolling a 1, 2, 3, 4 or 5 are the same as the chances of not rolling a 6, giving us 1 − 1/6 = 5/6.

Now think of a game that involves rolling a die several times. What are the chances, say, of getting two sixes in a row? Now we want to "roll a 6 and another 6," or "roll a 6 and roll a 6." There are six ways the first roll can go, and only one helps us towards our goal; even after that, the second roll gives us another six possible outcomes, and again only one of them will do. We must run the gauntlet twice, with a 1/6 chance of succeeding each time. We can capture this mathematically by multiplying the two probabilities: 1/6 × 1/6 = 1/36, which makes this a pretty unlikely event.

This combination of "or," "and" and "not" is extremely powerful (see De Morgan's Laws, page 126). With just these techniques you can, with a little cunning, calculate some really quite complicated probabilities. Here's a fairly challenging one: if you're dealt five random cards from a standard deck of 52, what are the chances of getting one pair whose values match (a pair of queens, for example) and no other matching cards? This sort of question is of considerable interest to poker players.

When a set of outcomes all have equal probability, we have the Uniform Distribution. It's a good starting point for the study of probability.

The Gambler's Ruin

Eventually the house always wins: this equation explains why.

Chance of Alan winning each round

Alan's starting funds

Chance of Alan bankrupting Betty

$$P(A) = \frac{1 - \left(\frac{1-p}{p}\right)^{f}}{1 - \left(\frac{1-p}{p}\right)^{t}}$$

Total starting funds

What's It About?

The French mathematician and philosopher Blaise Pascal posed a now-famous problem to his collaborator Pierre de Fermat in a letter of 1656; here's a standard modern version. Imagine Alan and Betty are playing a gambling game. Each starts with a certain number of tokens (perhaps not equal). Each turn they throw some dice and look for some predefined outcome — for example, "throw three dice and see whether at least one of them shows a 6." If it happens, Alan wins and gets one of Betty's tokens; if not, Betty wins and takes one of Alan's. They repeat this until one of them has all the tokens and the other is "ruined." The question is: what are the chances of ruin for each player?

The analysis of this problem led to a formula that yields surprising results. If Alan has a lot more money than Betty, and so can continue to play as long as necessary, Betty will almost certainly be ruined as long Alan has a tiny edge — that is, the probability of the outcome that's favorable to him is just a little better than ½. This is the principle on which almost all casino games are based: the house has a huge reserve of money and a small advantage. This is enough to ensure that, no matter how many gamblers win, the house eventually takes everything.

When the zero or double zero comes up in roulette, the house wins. This rarely happens, but tiny advantages like this keep the casino in business.

In More Detail

The game is a Markov Process (see Brownian Motion, page 70) and our equation can be used at each round to calculate P(A), Alan's chances of eventually winning, as if the game had just started with the tokens distributed as they are at that moment. If Alan has all the tokens at that point, we have P(A) = 1: after all, Betty's now ruined and the game ends. Similarly, if Alan currently has no tokens the equation gives us P(A) = 0; he's already bankrupt, so his chances of winning are nil.

Let's see how the funds each player has, and the presence of a slight edge in Alan's favor, affect the game. The slight edge means that $p > 1 - p$, which means

$$\frac{1-p}{p} < 1$$

Even if Alan has more than he started with, he's still in trouble. Betty probably has enough in the bank to ruin him.

That in turn means that this number raised to a big power is closer to zero than when it's raised to a smaller power. Let's look at the numbers when Alan has 100 tokens, Betty has 10 and Alan's chance of winning is $^{51}/_{100}$:

$$P(A) = \frac{1-\left(\frac{0.49}{0.51}\right)^{100}}{1-\left(\frac{0.49}{0.51}\right)^{110}} \approx 0.994$$

This is really just 1 minus something very small divided by 1 minus something just a tiny bit smaller, which is really very close to $^1/_1 = 1$. Perhaps, though, the conclusion is still amazing. If Alan opened a casino offering this game, allowed each player to buy a maximum of 10 tokens and had 100 of his own tokens on hand for each one, he would bankrupt almost every player: only about six

out of every 1,000 players would win the jackpot and take Alan's 100 tokens. Since he'd have won 9,940 tokens from the games he won, those six pay-outs shouldn't hurt too much.

Finally, note that the equation doesn't make sense for a perfectly fair game, because then the bottom of the fraction is zero, and dividing by zero is a Bad Thing. For that case we have the formula

$$P(A) = \frac{f}{t}$$

which simply says that Alan's chances of winning are equal to the proportion of the total pot he has at that moment.

The more money you have, the more likely you are to win in the end, simply because you can stay in the game longer.

The Gambler's Ruin

Bayes's Theorem

How worried should you be if you get a positive result from a very reliable test for a rare disease?

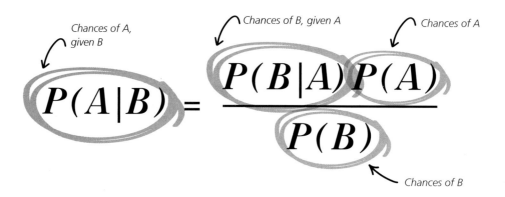

What's It About?

A doctor tests me for a disease that 1% of people with my symptoms have. The test is 99% accurate, meaning that when you test someone who has the disease it gives a positive result 99% of the time, and when you test someone who doesn't have it they get a negative result 99% of the time. How worried should I be if I get a positive test result back?

It seems I ought to be very worried; after all, the test is very accurate. So I go back to see the doctor, with much trepidation. The doc, who has some experience with these things, tells me to take the test again; I do, it comes up negative and to my delight I'm given the all-clear. What just happened?

Bayes's Theorem tells us that although the test is very good, the positive result only means I have a 50% chance of having the disease. Most people find this extraordinary. In fact we're very bad at making judgments about situations like this by gut instinct, which is one thing that makes Bayes's Theorem very useful.

This isn't to say that accuracy of tests isn't important. After all, the second test had only a small chance of being wrong, so when it came back negative we could feel fairly confident that I

really wasn't sick. The lack of symmetry between these two situations comes from the fact that having the disease is highly unlikely, but without Bayes's Theorem it's hard for us to weigh those different probabilities in the right way.

Why Does It Matter?

Bayes's Theorem is about judging the likelihood of something given that we know something else. This happens all the time, since in real life we usually have some relevant information. In 2012 statistician Nate Silver became a minor celebrity by correctly predicting the outcome of the U.S. general election in all 50 states and in the District of Columbia, claiming that Bayes's Theorem was central to his approach. In courts of law, Bayes's Theorem has been both used and abused in arguments connecting evidence to claims about guilt or innocence, in particular in relation to DNA evidence.

On a more mundane note, your email account probably has a spam filter that uses Bayes's Theorem. It starts off with a collection of common spam words and a set of probabilities of the form, "The probability that an email contains this word, given that it's spam, is X" and uses them to calculate the probability that the email is spam,

Known Unknowns: Chance and Uncertainty

given the words it contains. Over time the filter adjusts these probabilities and its collection of words based on the emails you receive. Similar techniques are used by linguists to develop software that can analyze, parse and reproduce texts written in natural languages.

The theorem is also used regularly by psychologists, pollsters, geneticists, physicists, hackers, linguists, company directors, military strategists and spies. It played historic roles in breaking the Enigma code and proving that smoking tobacco causes lung cancer. Even philosophers know about it. Actually, this brings us to something odd: perhaps uniquely among the equations in this book, Bayes's Theorem is controversial. The controversy doesn't arise from the correctness of the formula so much as how to interpret it and what we're really doing when we calculate probabilities. This leads down interesting philosophical avenues we can't pursue here (see The Law of Large Numbers, page 174), but it's worth being aware that this innocuous-looking equation has hidden depths.

Bayes's Theorem played a star role in the cracking of the Enigma code, making it one of the few equations in this book that can claim to have helped to defeat the Nazis.

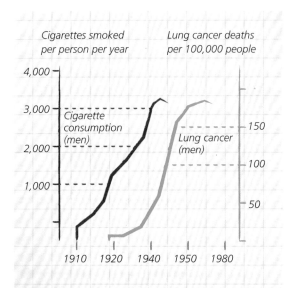

In 1951 Jerome Cornfield used Bayes's Theorem to argue for a probable causal link between smoking tobacco and lung cancer.

In More Detail

Sometimes our calculations of probabilities are affected by other things we know. If I rolled a die and hid the result, the chances of me getting a 6 would seem to be ⅙ (see The Uniform Distribution, page 162). What, though, if I tell you I got an even number? That certainly seems to make it more likely that I rolled a 6. So we ask: what are the chances that I've got a 6, given that you know I rolled an even number?

Perhaps you can make an intuitive guess, and this time you may well be right, but let's use Bayes's Theorem to calculate it step-by-step:

$$P(\text{six, if we know it's even})$$
$$= \frac{(P(\text{it's even, given that it's a 6})\,P(\text{it's a 6})}{P(\text{it's even})}$$

The probability that it's even given that it's a 6 is easy: it's 1, because six is always an even number! The probability that's it's a 6 is $\frac{1}{6}$, assuming the die is fair, and similarly the probability that it's even is $\frac{3}{6} = \frac{1}{2}$, since there are three even numbers on the die. So altogether we have, after a little fraction-wrangling,

$$P(\text{six, if we know it's even}) = \frac{1 \times \frac{1}{6}}{\frac{1}{2}} = \frac{1}{3}$$

More generally we write $P(A|B)$ for "the probability of A, given that we know B has happened." The study of problems involving this kind of thing is called "conditional probability," and Bayes's Theorem is one of its cornerstones.

There's a standard way to compute the conditional probability of A given B. If A has yet to happen and B already has, we're clearly interested in the probability of both A and B happening. But

How likely is it that this person is left-handed? That depends on what information we already have.

that isn't quite what we want, because it includes the probability of B happening; we already know that B has happened, so we should factor out that element of uncertainty. This leads to what amounts to a mathematical definition of $P(A|B)$:

Equation 1

$$P(A|B) = \frac{P(A \text{ and } B)}{P(B)}$$

Problems in many areas can arise by confusing $P(A|B)$ with $P(B|A)$ which, according to Equation 1, aren't the same. This mistake — known as the "fallacy of the transposed conditional" — is especially common when assessing forensic evidence in criminal cases. The probability of the accused being guilty, given the evidence, is not at all the same as the probability of the evidence being what it is, given that the accused is guilty. The second of those is usually much more likely, and mixing them up can lead to a wildly wrong estimate of how compelling the case is.

You may wonder what happens when two events have nothing much to do with each other. For example, what are the chances that the next person you meet will be left-handed given that it's a Tuesday? Perhaps the day of the week makes a difference, but most likely these two events are what statisticians call "independent," which means exactly what it sounds as if it means: whether you meet a left-handed person has nothing to do with whether it's Tuesday or not, and vice versa. Put mathematically, $P(A|B) = P(A)$ — that is, your chances of meeting a left-handed person given that it's Tuesday are just your chances of meeting a left-handed person; the "given that it's Tuesday" part is completely irrelevant.

It would be nice to get from the definition of conditional probability to Bayes's Theorem. By simply switching around the letters in the definition we have

Equation 2

$$P(B|A) = \frac{P(A \text{ and } B)}{P(A)}$$

Known Unknowns: Chance and Uncertainty

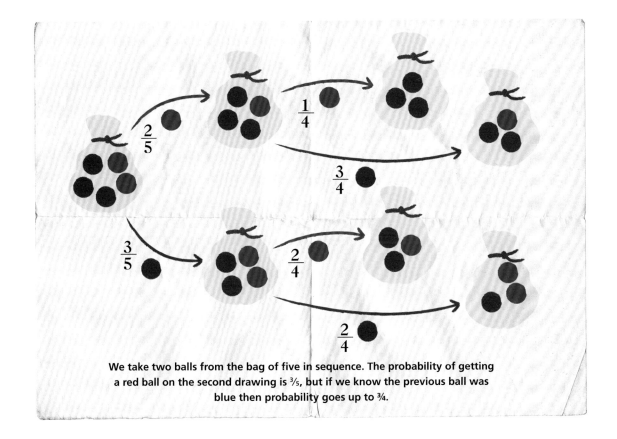

We take two balls from the bag of five in sequence. The probability of getting a red ball on the second drawing is ³⁄₅, but if we know the previous ball was blue then probability goes up to ¾.

What's more, by rearranging Equation 1 we can see that

$$P(A|B)P(B) = P(A \text{ and } B)$$

Substituting this into Equation 2 gives us the final result:

$$P(B|A) = \frac{P(A|B)P(B)}{P(A)}$$

So Bayes's Theorem really isn't mysterious — it results from the definition of conditional probability and a little bit of algebra. Yet is seems to be one of those basic facts most of us find very hard to grasp, which is why we often fail spectacularly at thinking clearly about probability. Next time you're told that some piece of information should change how likely you believe something is, ask yourself whether Bayes's Theorem is being applied correctly, wrongly or not at all.

If I know the chances of *A* given *B*, Bayes's Theorem can tell me the chances of *B* given *A*. But it depends crucially on other information about *A* and *B*, too.

Bayes's Theorem

The Exponential Distribution

**Wondering how long you can expect to wait for a bus,
a pay rise, a flu epidemic or an earthquake?**

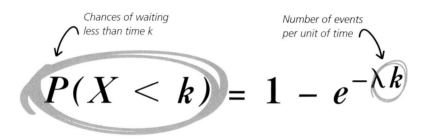

*Chances of waiting
less than time k*

*Number of events
per unit of time*

$$P(X < k) = 1 - e^{-\lambda k}$$

What's It About?

Imagine you arrive at a bus stop; a bus is supposed to come every 12 minutes, but we all know that in traffic they can get bunched up in odd ways, so we can't be sure when the next one will turn up. Still, they leave the garage (which is far away) on that kind of schedule, so we can reasonably expect to get about five per hour. What are my chances

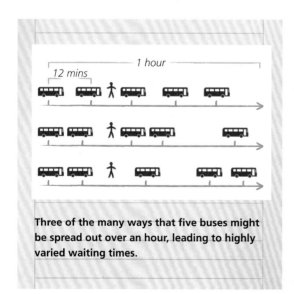

Three of the many ways that five buses might be spread out over an hour, leading to highly varied waiting times.

of a bus coming within the next 5 minutes? The Exponential Distribution gives us the answer.

The parameter λ is the rate at which buses come: in our case $\lambda = {}^1/_{12}$ because, peculiar as it sounds, on average one-twelfth of a bus comes every minute. The value k is a number of minutes, and the equation tells us how to calculate the probability that within k minutes we'll have caught a bus. For example, we have about a 34% chance of having got one within 5 minutes and — perhaps surprisingly — only a 63% chance of having got one after 12 minutes.

Now, suppose you wait for 10 minutes and no bus turns up. What's the chance that it will come within the next 5 minutes? It's 34% again. In other words, it makes no difference how long you've waited, your chances of having to wait another 5 minutes are exactly the same as they were when you first turned up.

In More Detail

Suppose I'm waiting at the bus stop and you're watching me from a window. You have better things to do than sit there staring at me, so you just check every, say, 2 minutes to see whether

Known Unknowns: Chance and Uncertainty

I'm still there. This will give you an estimate of my actual waiting time at the bus stop, accurate to within 2 minutes. It turns out that the chance that you'll have to check n times before I'm gone — and hence estimate I waited about $2n$ minutes — is given by the so-called Poisson Distribution:

$$P(X = n) = \frac{\lambda^{n}}{n!}\, e^{-\lambda}$$

Here, λ is just the number of buses that come, on average, every 2 minutes. Again, in our example this will be a fraction of a bus, this time one-sixth.

Now, suppose you check every 1 minute: this will make your estimate of my waiting time a bit more accurate. In theory, at least, you could

The Exponential Distribution is used to help predict rarer and more catastrophic events than the arrival of a bus.

continue refining it in this way: check every 30 seconds, every 10 seconds, every second, every 0.1 seconds … What happens when you do this is that your estimate of my waiting time gets closer and closer to the actual, correct value.

So, what is that actual value? It's given by finding the limit of that process as the time between peeps through your window tends toward zero (see Zeno's Dichotomy, page 18). All this means is that instead of repeatedly checking we'll continuously look out of the window: this gives us the Exponential Distribution.

**A neat equation that captures the probability
of waiting a given amount of time for an
uncertain event to happen.**

The Exponential Distribution

The Law of Large Numbers

We might get lucky once or twice, but in the long run do we always revert to the mean?

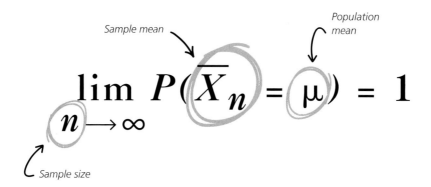

Sample mean

Population mean

$$\lim_{n \to \infty} P(\overline{X}_n = \mu) = 1$$

Sample size

What's It About?

Suppose we'd like to find out the average amount of money an American has in her or his pockets on a given day. One way to do this is to pick a small group of people — a sample — and have everyone put their money in the middle of the table. We then share out the pile equally. At that point each person has the same amount of money (barring a few stray pennies, perhaps) and that amount is the mean for the group.

In general, to find the mean of any set of data we do the same thing: add up all the individual amounts and then "share the total out equally" by dividing by the number of individual amounts we started with. This is one of the simplest and most common statistics used in science, business, politics and many other fields.

What we've calculated here is the "sample mean" — the average amount those people had in their pockets, not the average for the whole country. Intuitively, if we include more people in the experiment, we think that will make our estimate better, but will it really? The Law of Large Numbers reassures us that as we expand our sample the sample mean should approach a limit (see Zeno's Dichotomy, page 18) that is the true "population" mean — that is, a bigger sample gives us a better chance of our estimate being a good one.

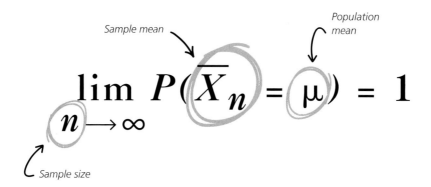

Average dice value against number of rolls

The average value we got when throwing a die many times gets close to the population mean of 3.5.

In More Detail

Imagine extending the experiment gradually to cover everyone in the United States. Clearly each person we add to the sample has a pretty small effect overall. What's more, there are a

Known Unknowns: Chance and Uncertainty

finite number of people in the population, so with enough patience we could gradually add to our sample until it includes all of them. At that point the sample mean will be the same as the population mean, because the sample really is the whole population.

The trouble comes when the population is infinite, or at least unlimited, as when the things we're averaging are spread out in time. Consider a gambler at the roulette table. The gambler knows the probability of getting red on any given spin is close to ½ (it's actually a little less [see The Gambler's Ruin, page 166]). So for any sequence of reds and blacks, we expect about half of them to be red and half to be black. In a short sequence, that might not happen: the chance of getting, say, three reds in a row isn't especially low, so we shouldn't be surprised to see that happen once in a while. If you watch the game long enough, though, you should see the Law of Large Numbers coming into effect and the sample mean of reds coming out at about ½. This is because, as the equation says, as the number of spins increases, the sample mean is very likely to approach the population mean.

Our gambler must be wary, though, of the "gambler's fallacy." If three reds come up, the Law of Large Numbers does not make black more likely next time. Each spin of the wheel is independent, and the chances of another red are still (about) ½. The law says that over many spins the sample mean ought to gradually close in on ½, but it says nothing about individual events or whether it might

The theory of probability, including the discovery of the Law of Large Numbers, were motivated by the 17th-century craze for gambling.

be a long way off that average for a long time before it reverts to the mean. Thinking otherwise belongs with winning streaks and lucky underwear in the mythology of gambling.

You might wonder, though, why there should be a true "population mean" for spins of roulette wheels when we don't know how many times the wheel will be spun. Does the population mean "exist" in this case? When the word "exist" gets put in scary quotes you know you're dealing with a philosophical question. All statisticians accept the Law of Large Numbers as a mathematical fact, but they can and do disagree as to what this equation really means.

The more times you try something chancy, the closer your overall results will be to the average.

The Law of Large Numbers

The Normal Distribution

From Napoleonic bureaucracy to credit derivative pricing, the Normal Distribution reigns supreme.

Chances X is between a and b

Mean

$$P(a \le X \le b) = \frac{1}{\sigma\sqrt{2\pi}} \int_a^b e^{-\frac{(x-\mu)^2}{2\sigma^2}} dx$$

Standard deviation

What's It About?

Suppose you're a Prussian general in the 18th century and you'd like to know how fit your population is in case you need to draft them into the army. You grab a sample of men of fighting age and take some measurements: height, weight, how much they can bench press and so on (I'm not sure about that last one, but you get the idea). Let's focus on height: what we probably expect is that a lot of people are close to the (mean) average height (see The Law of Large Numbers, page 174); a few people are a bit on the short side and about the same number are a bit on the tall side; and relatively rarely we'll measure people who are extremely tall or short. In other words, we expect the data to be shaped like a sort of bump, with most of the measurements being close to the middle and not many lying far away to either side.

The Normal Distribution formalizes this intuitive idea. Think of height as a random variable: that is, something that varies unpredictably when you pick random people from your population. A distribution describes how it varies, and enables us to calculate probabilities. We can then answer questions like "What are the chances a random person is over 1.8 m (6 ft) tall?", "How likely is it that a random person is within 10 cm (4 in) of the average height?" and "Is this person any good with a musket?" (actually maybe not the last one).

Why Does It Matter?

The word "statistics" comes from "state." Its study originates in the efforts of princes and bureaucrats to answer questions about their population's health, productivity and so on. Napoleon made it central to his project and ever since then, governments have collected and analyzed data about the people they serve. Very often, this data falls — or appears to fall — into a Normal Distribution.

In fact, though, the Normal Distribution has a very different origin. Astronomers had long realized that there was a chance of error with any observation, but they began to think about it more precisely in the 18th century, with the new science of probability at their disposal. Probably the error wouldn't be huge, but they often couldn't tell whether the measurement they'd made was a bit on the high side or a bit on the low side. Making multiple observations, especially when different people and equipment are involved, obviously helps to rule out certain specific causes of errors, but they can't be eliminated entirely. The astronomers came

Known Unknowns: Chance and Uncertainty

to expect that the data would be, as we now say, normally distributed around a mean value that was probably the actual value they were all trying to observe. Normal Distribution gives a way to factor out errors, although that's not guaranteed to work — there might, for example, be a bias that causes everyone's measurements to be too high, and then the average will be too high as well.

This so-called "normal law of errors" was taken over from science into a great diversity of subjects. Demographics we've already mentioned; soon the Normal Distribution was being used across the natural and social sciences and in other domains.

It's important to remember that in many of these cases it's acting as an approximate model. All too often the data are fitted to the Normal Distribution curve because it seems that's how it ought to be. This is explicitly done in some standardized tests, where "grading on a curve" forces the results to be normally distributed, regardless of the strengths or weaknesses of any particular cohort of students. In others cases, unconscious bias can cause people grading tests to produce something close to a Normal Distribution without meaning to, because they've come to believe that it "feels right."

On a more somber note, lifespan is assumed to be normally distributed: there's an average age we all make it to, some live longer and some less so. Sadly this isn't true, because some people die very much younger than the average because of illness, accidents, violence and so on. What do we do about this, statistically speaking? Why, we label those that throw our data off the normal distribution as "premature deaths" and deal with them separately.

In More Detail

The equation for the Normal Distribution is pretty weird-looking, and it's like that because it needs to satisfy some quite weird criteria. A distribution like this isn't used to work out the probability of getting a certain measurement because it turns out that the probability of someone being exactly, say, 6 ft tall is easy to calculate: it's zero. This may seem very strange, but notice that we really do mean exactly 6 ft, not 6.1, not 6.01 and not even 6.000000001. Well, we can't make measurements with infinite accuracy, so really the question isn't very sensible. Instead, we should ask something like, "What are the chances that someone will be between 5.9 and 6.1 ft tall?" This is how we always have to work when we're measuring a continuous property such as height rather than counting discrete things such as children.

Sometimes we massage our data to fit the Normal Distribution because otherwise it seems wrong.

The Normal Distribution

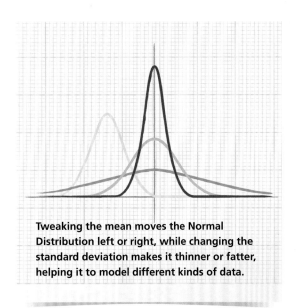

Tweaking the mean moves the Normal Distribution left or right, while changing the standard deviation makes it thinner or fatter, helping it to model different kinds of data.

To see how the Normal Distribution helps answer that question, we plot a graph of it, with the x-axis being the height (say) and the y-axis being the value of the distribution given by the equation. It comes up in a reassuring bump with the summit lining up with the mean height. We now mark off the heights we're interested in and use a spot of calculus (see The Fundamental Theorem of

Calculus, page 26) to find the area under that part of the curve, which is the probability that the height of a random person will fall in that range. This means the total area under the curve must equal 1, because that's the probability that a random person has any height whatsoever. In other words, it's certain that they will have some height. In general, though, the Normal Distribution doesn't have maximum and minimum values: it allows very extreme values indeed, both positive and negative. It just makes them very improbable. This is one of many reasons why heights are not really normally distributed: nobody is 3 m (10 ft) tall, and nobody has zero or negative height. Close to the mean, though, it might be an OK approximation. This is partly why the definition's so convoluted: it's a curve that's infinitely long in both directions, never touches the x-axis, but the total area under it is exactly 1.

There are many other probability distributions, but this one has a special place in statistics because it arises from the others in a special way. Imagine you're tossing a coin. Whether you get heads or tails is uniformly distributed (see The Uniform Distribution, page 162). Now, though, let's play a game: toss the coin 10 times and count the number of heads. The result of an experiment like

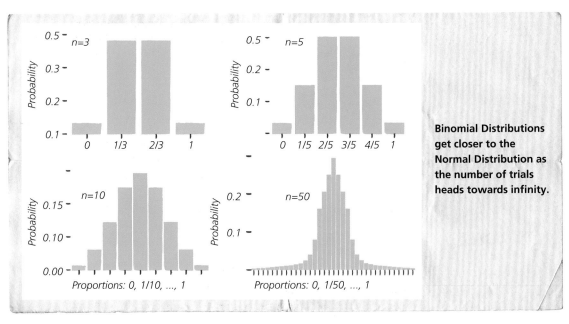

Binomial Distributions get closer to the Normal Distribution as the number of trials heads towards infinity.

Known Unknowns: Chance and Uncertainty

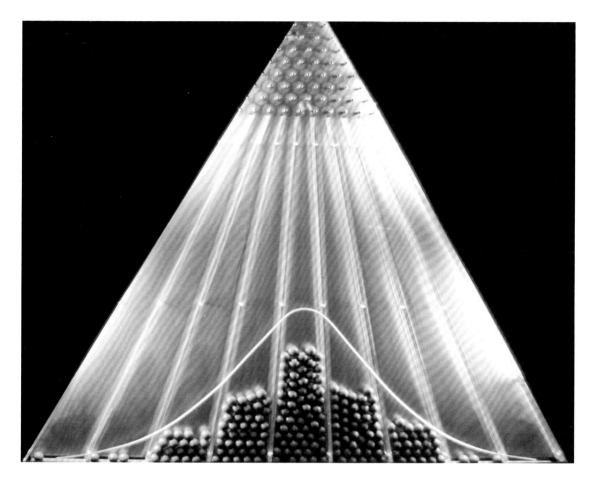

The quincunx is a mechanical device that demonstrates one way the Binomial Distribution can arise in simple physical situations.

this is governed by something called the Binomial Distribution. If you do this lots of times you'll find that the mean is 5 and the other scores fall away on either side in a way that looks suspiciously familiar. Now try it again, but this time toss the coin a hundred times, and again do this many times. The result will almost certainly look even more like the Normal Distribution. In fact, something called the Central Limit Theorem says that the more coin tosses you do in this way, the closer your results will get to the famous bell curve. Notice how remarkable this is: tossing coins seems to have

nothing at all to do with Normal Distributions. Although this is a rather technical result, the slogan is easy enough: the Normal Distribution pulls many other random variables toward itself, like a sort of statistical black hole, when we repeat our measurements many times over, as long as the measurements are independent of each other.

This strange-looking equation offers a useful model of many phenomena, although it's also easy to abuse.

The Normal Distribution

The Chi-Square Test

A common way to test whether or not your probability distribution fits your data.

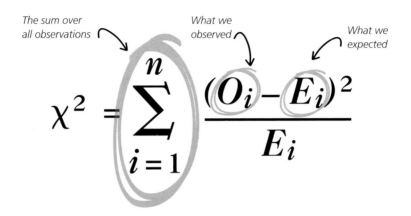

$$\chi^2 = \sum_{i=1}^{n} \frac{(O_i - E_i)^2}{E_i}$$

The sum over all observations — *What we observed* — *What we expected*

What's It About?

We're playing a board game, but I'm afraid I suspect you're cheating — I think you might be using a loaded die. If not, the distribution should be uniform (see The Uniform Distribution, page 162), meaning every number has an equal chance of coming up. That doesn't, though, mean that every number will have come up an equal number of times during our game, because you may have been lucky today even if the die is fair (see The Law of Large Numbers, page 174). I ask to swap dice, but you take offence and ask what grounds I have for suspecting you of such a vile deed — perhaps the reaction of a guilty person, but I'll need something better than armchair psychology to resolve this one.

Now's a good time to carry out a quick Chi-Square Test that can, among other things, give us a rough idea of how likely it is that the die is biased, based on the rolls you've made during the game (of which I have, naturally, kept a scrupulous record).

In More Detail

The idea is to compare the observed frequency of each possible outcome with the frequency we expect to see if the distribution is what it should

be. We then sum these up to get a number. The number on its own is pretty meaningless, but we can use this to look up a number that tells us roughly how confident we can be that the distribution of the die's rolls really is what we think it should be. .

Let's use an example. Suppose you've rolled 6, 3, 4, 4, 6, 1, 5, 2, 1, 6, 6, 6. That last 6 was the final straw: you've rolled far too many of them,

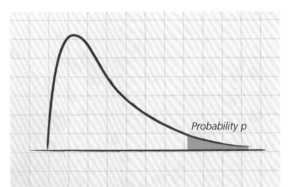

Probability p

The Chi-Square Distribution. The area under the whole curve is equal to 1; the probability that our observations don't fit the data is the shaded area on the right.

Known Unknowns: Chance and Uncertainty

I think, for this to be a fair die. What we want to do is test my hypothesis that the die doesn't follow the Uniform Distribution. Standing against this we have what's called a "null hypothesis," which is the ordinary, boring case that the die is fair after all. The Chi-Square Test allows us to generate a number called a p-value, which in turn gives us an approximate level of confidence that the die is biased. All this is generally done with either printed tables or a computer.

So, let's check your die rolls. There are 12 rolls, so if the die was perfectly fair we'd expect to get each possible number twice. Plugging in the number of actual observations of each score (O^i) and comparing them with the expected number $E^i = 2$ gives us, in order of score from 1 to 6,

$$\chi^2 = \frac{(2-2)^2}{2} + \frac{(1-2)^2}{2} + \frac{(1-2)^2}{2}$$

$$+ \frac{(2-2)^2}{2} + \frac{(1-2)^2}{2} + \frac{(5-2)^2}{2}$$

$$= \frac{1}{2} + \frac{1}{2} + 0 + \frac{1}{2} + \frac{9}{2} = 6$$

A total score of 6, then, but what does it mean? To find out, we look up the chi-square value of 6 in a table and it gives the probability that the die was indeed following the Uniform Distribution. In our case the answer is that we have no good reason to reject the null hypothesis, and I owe you an apology.

This kind of statistical hypothesis-testing is extremely widespread. It's far from perfect. You might very well be using a loaded die; all the

The Chi-Square Test gives an indication of how likely it is that your data really fit the distribution you think they do.

test says is that you could quite easily have got that combination of rolls with a fair one and I shouldn't jump to conclusions. Still, a lot of statistical arguments are like this: they're designed to appeal to common sense and similar standards of rationality rather than the rigors of pure mathematics.

This equation measures how well your data fits a particular distribution by adding up all the ways the fit isn't perfect.

The Chi-Square Test

The Secretary Problem

How many people should you interview before you pick one to hire?

Chances that the best one after x have been seen is the best overall

$$P(x) = -x \ln(x)$$

What's It About?

Some people are terribly indecisive. Perhaps you've been out with one of them and had to pick a restaurant to eat in. You walk along a few streets checking out the options. Some are obviously better than others. You're tempted by this nice-looking French bistro, but your friend wants to keep going. What if there's a really great place just around the corner? Despite your growling stomach you admit that maybe there is, and so you trudge on. The two of you are agreed on one thing: you'll settle on a place when you come to it, not retrace your steps to one you've already rejected. Who has time for that?

In fact, many decisions have to be made this way. One such is hiring someone for a job: you could have an almost endless number of applicants, but you'll have to make a decision before you've seen all of them. In fact, each one you see has a cost to you in employees' time and in the delay in filling the post. Admittedly this scenario — which gives the problem its best-known name — isn't quite like the restaurant one, because you can usually delay offering a job until you've interviewed a bunch of candidates and picked your favorite. To simplify things we assume you have to hire

the person you've picked on the spot — or alternatively, pick a restaurant when you come to it without backtracking to a previous one. So the question is: how many should you turn away before choosing the next good one you see?

Why Does It Matter?

Decision-making can be hard, and while deciding where to eat isn't a matter of life or death, in other contexts making the wrong choice often has a considerable cost attached to it. In business, the cost is usually financial; in a political or military setting it might really be life and limb at stake.

Wherever possible, mathematicians like to help people make the best decisions they can. The only way to be absolutely positive you've made the right choice is to check every possibility and compare them all. If that's not possible, can we do better than just choosing at random or giving up when we feel exhausted?

I touted probability as a way of dealing with uncertainty, but in a sense it relies on certainties that are one level removed. For example, I don't know what number will come up on my next die roll, but I know I have a 1/6 probability of getting a 5 (see The Uniform Distribution, page 162). This

Known Unknowns: Chance and Uncertainty

is because ordinary dice are deliberately designed to have a uniform chance of landing on any of their six sides. Away from the gaming table and the probability classroom, though, life isn't always like that. Sometimes it's not just that we don't know what will happen, we don't even know what the probabilities are of the different outcomes. How likely is it that this restaurant is the best one in town? Unless we have some external information we seem to be in a hopeless situation. Yet we must make a choice, and occasionally such choices pertain to even more important matters than getting a good dinner.

Decision theory sits somewhere on the overlap of probability, game theory, logic, sociology and psychology. It tries to formalize the ways we make choices and assess them; it also aims to show us ways we can make better choices, given our stated preferences. As a field, it's important everywhere where strategic decisions must be made and justified and there's something significant at stake. The Secretary Problem, humble as it is, is just one

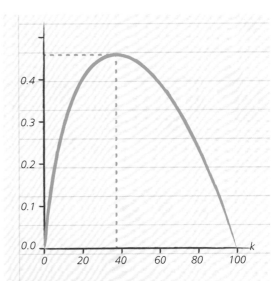

As you walk along the road (horizontal), the probability that you should pick the next restaurant you see that beats the others rises. It reaches a peak when you're $1/e$ of the way along the road and then declines.

Trading off risk for reward is an everyday feature in business and military affairs as well as in simple games like rummy. Decision theory aims to help us think through such situations.

The Secretary Problem

example of the surprising results that can come from mathematically analyzing the process of making a certain kind of decision.

The way we do this is often to develop a precise strategy for acting: a set of clear-cut instructions that replace the emotional tugs, frustration and dithering that beset ordinary choice-making. Indeed, these instructions usually take the form of an algorithm, a procedure so unambiguous you could turn it into a computer program and have a robot carry it out.

In More Detail

The aim of the game in the Secretary Problem is not to *definitely* make the best possible choice. At least, if it is, there's only one way to do it: check every option. My friend and I can't be sure there isn't an amazing restaurant over on the other side of town that only a thorough search will throw up. Searching that far afield is impractical, though,

so we're looking for a strategy to check as few as possible but maximize our chances of having included the best one. We'll look for a certain amount of time to get a sense of what quality of options is available around here, then we'll pick the next one that's better than any we've seen so far. All that sounds pretty vague, and it is, which is why it's hard to think about, so let's formalize it a bit and see if we can turn it into an algorithm.

Let's stick with choosing a restaurant. We'll assume that the restaurants on a street aren't lined up in order from best to worst, but are mixed up; otherwise the problem's too easy. Suppose there are n restaurants in total to choose from. We'll take a random sample of them represented by a fraction of the whole — so if we're going to reject half of them, $s = 1/2$. The sample is the proportion we'll see first and reject, after which we've "seen enough." The next one that's better than any of those we've seen so far is the one we pick.

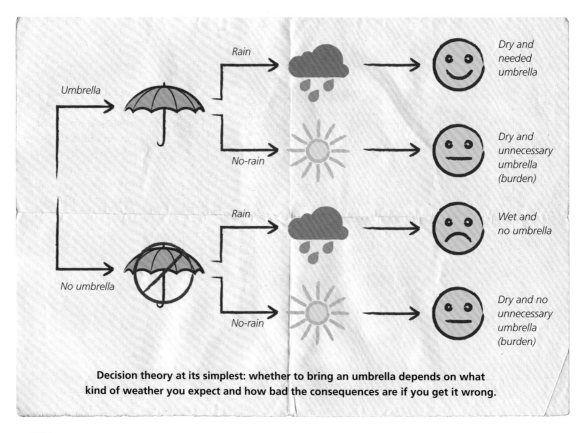

Decision theory at its simplest: whether to bring an umbrella depends on what kind of weather you expect and how bad the consequences are if you get it wrong.

Known Unknowns: Chance and Uncertainty

This is a good algorithm already, except for one thing. We need to know how big that fraction should be. If we don't see enough, there's a chance we'll settle for a below-par option. If we see too many, there's a chance we'll reject the best one and, again, end up making a less good choice. Of course, with any strategy there's a chance we won't get the best result, so we're looking for the strategy that minimizes that chance, making it as likely as possible that we'll get the best of the lot.

The actual calculation isn't especially difficult, but it's also not very enlightening. It uses conditional probability (see Bayes's Theorem, page 168) to find an expression for the probability, for a given fraction of the total number, that the best candidate isn't in that fraction but the second best candidate is. This guarantees that after rejecting the sample (including the second best candidate)

we'll only settle for the best candidate, whom we haven't rejected yet. The surprise is that the best choice of fraction, s, is the one for which this limit is true:

$$\lim_{n \to \infty} s = \frac{1}{e}$$

where e represents the base of natural logarithms (see Logarithms, page 36). Your calculator will tell you this number is about 0.3679. That means that after you've squinted through the windows of about 37% of the restaurants, or interviewed about 37% of your job applicants, you should choose the next one that's better than any of the ones you've seen so far. No guarantees, but this strategy gives you the best shot at ending up with the finest meal and the best employee you possibly can.

Though the interviewee may feel nervous, the interviewer should be worried too: it's no easy matter to make the right choice when you can't see every possible candidate for the job.

The application of math to decision-making is fraught with difficulties but can be very valuable. This famous example gives a surprising answer to an apparently simple problem.

The Secretary Problem

Index

Secret Language of Equations

Secret Language of Equations

Secret Language of Equations

Author's Acknowledgments

The author would like to thank Clare Churly, who came up with many ideas he passes off as his own herein; Robert Kingham, for eagle-eyed reading and everything else; Nathan Charlton and Andrew McGettigan for making him think and then do something about it; and of course the students and staff at Central Saint Martins, City Lit, Queen Mary and elsewhere.

Editorial Director *Trevor Davies*
Senior Editor *Alex Stetter*
Copy Editor *Caroline Taggart*
Art Director *Jonathan Christie*
Designer *Tracy Killick*
Illustrators *Emily@kja-artists.com* and
  Peter Liddiard at suddenimpactmedia.co.uk
Picture Researchers *Giulia Hetherington,*
  Jennifer Veall
Assistant Production Manager *Lucy Carter*